flirtexting

flirtexting

how to text your way into his heart

Olivia Baniuszewicz **and** Debra Goldstein

Skyhorse Publishing

www.skyhorsepublishing.com

10 9 8 7 6 5 4 3

Library of Congress Cataloging-in-Publication Data

Baniuszewicz, Olivia.
  Flirtexting : how to text your way into his heart / Olivia Baniuszewicz and Debra Goldstein.
      p. cm.
  Includes bibliographical references and index.
  ISBN 978-1-60239-367-7 (alk. paper)
 1. Man-woman relationships. 2. Flirting. 3. Text messages (Telephone systems) 4. Cellular telephones—Social aspects.
I. Goldstein, Debra. II. Title.
  HQ801.B17 2009
  306.730285'4692—dc22
                                        2008045959

Printed in China

To my Dad, who set the bar so high no one's been able to reach it.
—Debra

To my Mom, whose advice will always be the first and last I'll take.
—Olivia

contents

Dear Reader,

It's official. Boys text; therefore girls need to know how to flirtext. As single, social gals in a big city, we recently began having convos over text that would normally happen over the phone. We realized that because this phenomenon was still new, there were no set rules or guidelines, and certainly no books about how to handle these flirtatious texts. Therefore, we felt the need to regulate. The two of us are relatable girls with relatable stories. We took it upon ourselves to continue dating off the chain and began keeping a record of the texting situations we encountered. We've created this user-friendly guide so that you can take what we've learned, apply it to your situation, and feel confident when pressing send.

Flirtexting is chock full of useful examples (text this, not that), guidelines for creating the perfect flirtext (what to say to get what you want), encouraging advice (say buh-bye to drunk texting), and explanations for decoding his text (what he really meant by that text). We highlight and address some classic texting situations that are sure to occur in the beginning stages of your relationship. This book is your guide to everything flirtext. We encourage you to highlight, dog-ear, and keep handy to your cell at all times.

We want to mention that for the most part our advice applies to the courting stages of a relationship (roughly the first six weeks or so). However, there is a great deal of advice throughout the book that will help you at any stage of your relationship, as well as a chapter dedicated to those wanting to spice up an existing one. Obviously, every relationship is situational, so we ask that you use your best judgment in deciding where yours falls. Keep in mind that flirtexting is an art not a science, much like the act of love.

We write to you as we would to our BFF. The advice we give is written with honesty and delivered with love. *Flirtexting* is a compilation of our personal experiences as well as those of some hot guys and dolls who graciously loaned their stories for our cause. We found it necessary to write this book so that girls like us have a place to turn to get all of their flirtexting questions answered.

Finally, a guide that demystifies the *flirtext*.

XOXO,
Deb and Liv

why text?

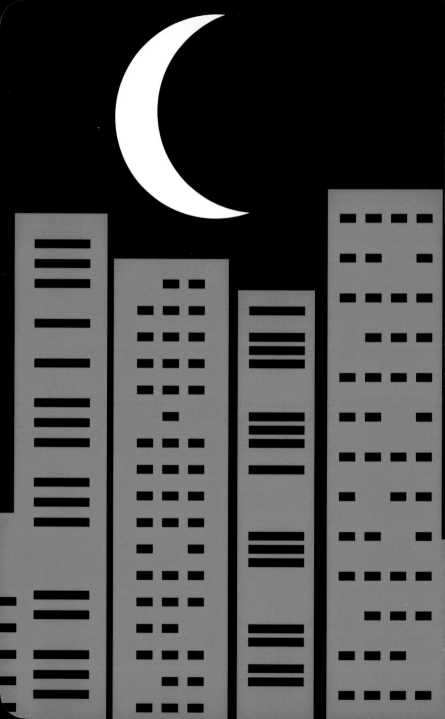

Introduction

Text in the City

"I'll do whatever it takes to get girls to go out with me. If the new thing is wearing a bag over my head, then that's what I'm gonna do. Right now it happens to be texting. So I'm all over it."—**Eric**

Recently, we were having drinks in the city with a few of our girlfriends when Olivia received a text message from her current crush. We instantly got excited and screamed for her to read it out loud for us all to hear. After analyzing every last word he wrote, we each shared our brilliant thoughts on how Olivia should respond. We put lots of time and energy into concocting something delicious … and finally came up with the best possible text! And as an added touch, we delayed sending the text for two hours. The result: an immediate response and, soon after, an official date invite. Believe it or not, you have the secret weapon in your hand to whatever kind of relationship you want with the guy on the other end—friendship, innocent flirtation, possible hook-up, or romance.

Your cell phone has become the most widely used instrument in scoring that hot date, resulting in the emancipation of *flirtexting* (or the *flirtext*).

> **A flirtext is any text message sent between you and a boy you would like to date or are currently dating (i.e., someone you would like to make out with or are currently making out with). These texts are normally comprised of the flirty, witty banter that you throw back and forth with each other that typically leads to a date and, if you're lucky, then some. You love this type of flirting over text because it's a noncommittal way for you to let him know you are digging him.**

We're single, independent, sassy gals in our mid-twenties who live in a big city and love to party, date, and enjoy being young. We're flirts addicted to the social scene, and we welcome being the center of attention.

At the start, we had the same questions (Does he really like me if he's texting and not calling?) and concerns (Is it too soon to text back?) as you. Through the surplus of single and suitable boys that our city has so graciously left on our doorstep, we've been

The term "text me" became the mating call of singles ready to mingle during our prime dating days.

able to play the field while using the logic of trial and error to create our method. We gained confidence with each and every text that we received and sent, leading to more date invites than days of the week.

Our outcome: Flirtexting has become the NEW "first step" in dating. It's time to upgrade to that unlimited texting plan and sharpen your typing skills. Dear reader, we are passing on tried-and-true information, perfected in the finest dating capital in the nation, New York City. Whether it's "Mr. Right" or just "Mr. Right Now" you're pining after, flirtexting will help you get what you want. In *Flirtexting*, we provide the ultimate guide to wooing your guy over text. After using the skills taught in this book, you will have your latest prospect in the palm of your hand—literally. Message received?

texting tell-all

See how we came across the beauty of the text and made it work for us.

"It's not a southern thing, it's a Deb thing."—Debra

Growing up in the South has preconditioned me to a dating etiquette that is as traditional as it gets. I expect boys to call me first, open my door, make the first move, and laugh at all my jokes (okay this may not be a Southern thing … it's a Deb thing). And when this doesn't happen, I can't say I'm not disappointed. Although I've become more forgiving toward boys as I've gotten older and moved out of the South, when it comes down to basic dating courtesy, I won't budge.

So you can imagine my frustration when boys began asking me out over text. I kept thinking to myself that *my* Mr. Right would never text message me out on a first date. Being on the slightly stubborn side, I held on strong for a while and turned down any guy who asked me out over text.

This went on for some time and consequently my dance card became less full. It wasn't until a cutie with true potential came along—a guy who, like most, just loves to text—that I broke this little rule of mine. And then I cut the next guy some slack, and so on and so forth.

Cut to … three years later … and we're writing a book about how great texting and dating is. 'Nough said.

texting tell-all

"I got asked to my prom via text message!" —Olivia

It was my junior year in high school. JD sent me a text one day after school. JD's ID? He was my friend and a hot senior boy whom sources had revealed had a mini crush on yours truly.

The texting went back and forth, as we reviewed each other's day and sports practices, which led to:
"Are you going to prom?"
"Yes, obviously."
"Do you want to go together? It'll be fun."
"Really? Over text JD?"
"Is that a yes?"

Even though I had seen JD about two hours earlier, he neglected to hold this convo face-to-face (F2F). Needless to say, he ended up escorting me to my much-hyped prom. Being a girl who loves traditional approaches, this invitation left me somewhat embarrassed and confused. After all, my date put as much thought into asking me as Pauly Shore put into *Bio-Dome*. I am pleased to report we rocked the prom despite that shaky start.

We Date, Therefore We Text

"It's the chance to demonstrate my quick mind and biting wit … and the possibility of hooking-up as a result of such brilliance." —Marc

Somewhere between the time when *90210* went off the air and then came back on (with a new cast), we came to terms with what we knew all along but weren't ready to admit: Texting has become the NEW first step in dating. Sadly, we had to say "buh-bye" to our traditional, and now outdated, way of thinking, and have learned to embrace flirtatious texts and the relationships that they breed. Upon discovering the undeniable power of texting, we finally put down our fighting gloves and our phone simultaneously became our BFF and our outlet to dating. We soon realized that texting really isn't all that bad. In fact, it's pretty wonderful.

This is a book about the *only* means of communication that has forever changed our views of the modern courting stages of dating. We know what you're thinking. What about e-mail and Facebook? Aren't they also ways of "communi-dating"? Yes, but this is not a book about them. As far as we're concerned, we live by only one reliable and capable means of techno-relating, and that's the text message. Moving on.

> **Communi-dating:** **How we date through our various forms of digital and wireless communication, i.e., text, BBM, e-mail, phone, Facebook, Match.com, etc.**

Our hectic schedules benefit from a sweet shortcut as often as possible. Between waking up, having a productive day's work, hitting yoga, eating dinner, watching *The Office,* and going to bed, it's difficult to reconnect with those near and dear to us. Thankfully, technology has been able to keep up with our busy lives. Now, if we don't have time to call someone, no big deal; we can text. Texting takes ¼ the amount of time of talking, and allows us to connect with more people than we would with a phone call.

Perhaps the name explains it all: SMS stands for Short Message Service and speaks volumes about our generation; one that likes to keep it short, sweet, and to the point. UG8TIT? When given the opportunity, we'd so much rather say it in four words or less. Look at it this way: If

> *The text is our new-age electronic Post-it note, just without the sticky part!*

you're given a large piece of paper to write a short message, you will feel pressure to write a lot and divulge unnecessary details. But if you're given a Post-it note to write that same message, you'll write only what you intended—*a short message.* Plus, with ADD (Attention Deficit Disorder) affecting 99 percent of our peers (or so it seems), text allows us to get straight to the point: "Fro yo vs ice cream throwdown—are you game?"

"Hi. Want to grab a drink 2nite?"

Um, hello! Is this supposed to be charming? If he really cared, wouldn't he call or ask me out in person? Have you turned down tons of decent boys just because they were texting and not calling? Well, so did we, gals ... so did we. Until now! We stopped fighting it and figured out how to make flirtexting work for us, and so can you.

texting tip

THE BOTTOM LINE:

When done strategically, texting can help you get a date and potentially a boyfriend. Exactly what you write back and when you write back will determine which direction your relationship will go.

Flirtexting is an efficient means for guys and girls to explore one another's feelings and see if there's mutual attraction. If the response is positive, and you both like what you read, you'll eventually take the next steps: Facebook friends, phone call, date, meet the parents, adopt a few kids, the usual. To get to that point, you must first impress each other through text. This book will give you the skills you need in order to feel prepared and confident when pressing send. And if you play your texts right (do what we say), you'll ultimately learn how to seduce the man of your dreams, *through text*.

Our motivation for writing this book came from our hot girlfriends' overzealous efforts to hookup with their Prince Charmings, fall in love, and get married. Single women everywhere in their twenties and thirties have worried about this for ages. Societal pressures to get married and start a family

> *Beware of being "fast track Jane": In due time Mr. Right will find you. Don't be "Jane," desperate to find love.*

are still very present today, leading to the notion that there must be something wrong with you if you don't settle down by a certain age. But we say: Chillax! Understand that your time will come, and realize that falling in *true* love cannot be forced. And more often than not, it arrives unexpectedly. Therefore, it's time to take a chill pill (literally, if psychiatric need be), work on ourselves first, and then allow Mr. Right to find us. Trust, that in due time, he will. It's inevitable.

> **Fast Track Jane:** The insecure girl (in all of us), who at times lets the overbearing societal pressures of getting married get the best of her. This "fast track"she is pushed to be on causes her to act in unflattering ways, that can turn guys off.

When working on our relationships, we can sometimes lose sight of ourselves. The fight for love has led girls to move cross-country to follow their latest prospect, to act/dress/speak like an ex-girl-friend, take an interest in deep-sea fishing (despite being deathly afraid of water), and even turn Republican. Undoubtedly, the dedication to finding love is there; however, the pressure to get that ring and settle down can become so overpowering that sometimes girls don't realize that their attempt makes them come across looking transparently desperate.

texting tell-all

Deep thoughts by Deb and Liv:

There's a large misconception that we need to be with a partner in order to be happy. This innate pressure many of us carry has led girls to desperately seek someone, *anyone* in the name of love. Well we've got news. "Desperate" is not a look you want to be sporting when searching for love. The best advice we've ever received was to become the person we want to marry. Ultimately, you're the only one that can make yourself happy. Strive to have all the qualities you look for in a partner so that you are complete without needing to depend on another person. Don't look for someone to fill a void in yourself. The goal is not to complete each other, but rather to complement each other so that you can grow together. Work on being happy with yourself first. Then your relationships will flourish, because you have stopped depending on someone else for your happiness. This is the key to a happy you and a lasting relationship.

We've seen the gorgeous and talented get rejected, the sweet and smart turn needy. We've watched the independent and fun suddenly turn dependent and pushy, and then get dumped ... at the altar. In due time, the women who attracted these men in the first place disappear, and the men shortly thereafter. Sadly, these girls more often become the rejectees and not the rejectors. And knowing that these women really are *quite the catch*, we don't like what we see.

It saddens us to see our beautiful, smart, remarkably witty girlfriends scare off guys due to their unconscious plays to get them to commit. Therefore, we found it necessary to eradicate any desperate girl syndrome by teaching confidence through text. There's no better way to hook a guy than through a witty flirtext. Texting has invaded our lives and our little black books. We've embraced it and it's done wonders to our dating lives, and now it's your turn! Consider this our gift to you: intervention for love. Ladies, allow us show you how to text your way to love.

Authors' Note:

We would like to encompass all the girls we referenced above, and the societal fast tracks they're pressured to be on, throughout the book. We'll reference them as a character named Jane. Jane is desperate to find love. Jane is every girl's faulty side when she's playing the field of love. Don't be Jane.

texting tip

As communication majors in college,

we learned that, invariably, the way a message is sent is more important than the content of the message itself. Due to its informality, texting has allowed us to become more approachable and open with our feelings toward one another. The low-risk, no-strings-attached approach that texting provides has given us a newfound confidence. A confidence that has ignited fire under our fingertips and hope in the eyes of the beholder.

It is important, however, to acknowledge Jane's presence, as it is vital in acknowledging our own faults and addressing them head-on. Jane is meant to be a reflection of our social imperfections and conundrums of dating. She will be your "what not to do" gal. Learn from Jane's mistakes. There are vital lessons to be learned.

SOME BACKGROUND ON TEXT

Each of us got our first cell phone in the late '90s, right around the time we began learning how to drive. Cell phones were new to the market back then, expensive, and mostly for "emergency" purposes only! Ha! If by "emergency" our parents meant calling every single one of our friends to and from school and in between classes, then the two of us were disaster prone. Our poor parents were stuck with sky-high phone bills while we were constantly grounded for going over our minutes. Ten years later, our emergency-purpose-only phone has now become our end all be all tool for communicating and, more importantly, communi-dating. We sleep next to it, check it first thing in the morning, and hold onto it for dear text. Next to television and PerezHilton.com, it's our primary access to the outside world and holds the key to our buzzing social lives.

Texting has come a long way since its introduction in the mid-80s as a tool for those who are hard of hearing to be able to communicate wirelessly. Since its invention, text messaging has become the most widely used mobile data service on the planet, with more than 363 billion texts sent in the United States alone in 2007. Wowzers!

SMS is the fastest form of communication ever invented. It sends and delivers messages faster than phone calls and voice mail. It's convenient, accessible, and free. Executives in the business world picked up on this and have made texting their most preferred method of communication worldwide. In fact, we just heard of someone making million dollar business deals over text. All reasons why anyone who hasn't already should jump on the text message bandwagon, and fast!

> **Reachability:** The addictive need to remain connected. A recent study revealed that text messaging is the most addictive digital service on mobile or Internet. Its addictive tendencies have been compared to the habit of smoking! Texters Anonymous, anyone?

LITTLE BLACK BOOK:

our phones and the lives in them …

The little black book used to be a small, usually black, notebook in which guys and dolls would store names, numbers, and notes of everyone they've dated, are dating, or would like to date. Frequently you would find little stats about each person next to their name. For instance, "Alex: from Chicago, likes football" or "Rich: Great kisser, big jerk" … you get the idea. How clever and handy.

With convenience and speed on its side, the cell phone has taken over the function of the little black book. You can still make those same notes as you would before, like "Dave: Met at Sarah's bday dinner. Sooo hot want to date!!" Only now, you can also use it to contact Dave with the touch of a button. This slick mini device stores not only our contacts, but our stories and secrets too. Plus, cell phones are digital so you can back them up and never have to worry about losing anyone's information. The little black book has gone wireless and our social calendars have blossomed because of it.

BENEFITS OF FLIRTEXTING:
let's recap

1. Outlet for mad flirting!

If you're anything like us, you speak in flirt! No matter who the fella is, it's part of your nature to tease, smile, and bat your eyelashes. Texting allows you to do this on a totally different level! You can flirt with many boys at once and only your cell will know about it. If you've got three boys you've been crushing on, text them all to see what they are doing and hope the one you like the most asks you out!

2. Helps eliminate social barriers such as shyness.

This is your chance to experiment with being more open about your feelings in a controlled environment. Text is taken a lot more lightly than a phone call, e-mail, or F2F. Therefore, it allows you to take more chances in life and in love. Use text as your testing ground to say things you might be hesitant to say in person. If it ever backfires, you can always just say that you were only kidding. Did he not hear the sarcasm in your text?

3. Girls can ask boys out ... and not feel weird about it.

In the olden days, or B.T. (before text), it was quite unusual for a gal to ask a boy out. Since texting wasn't around during *The Rules* conversation, no clear-cut rules were ever set. Text has created a loophole to our moms' old-fashioned approach of waiting for him to ask us out first, while keeping our code of etiquette above board. So if there's a cute young lad you're pining after ... text away!

4. Makes rejection easier to handle.

If a boy is calling you and you just aren't feeling it, texting him back instead of calling delivers the message, and avoids the awkward rejection.

That makes it easier on you both and will hopefully save the friendship. This also applies for the opposite of this scenario. If you text a boy and he doesn't write you back, aka

> *Being rejected over text is a lot easier to handle than being rejected in person or over the phone—or so we've been told;)*

dead air, then nine out of ten times, he's just not that into you. Ugh, the perils of dating.

5. *Fits into our fast-paced lives.*

Texting makes it easy to keep in touch in a quick and efficient way. When we don't have time to pick up the phone and call, a short text gets the job done fast and easy. While making a phone call wastes your crucial getting ready time, sending a quick text to let him know you're running late for your date is sufficient.

TEXTING BY DATE OF BIRTH

Texting is subjective. It is what you want it to be. Think of texting as the Op-Ed section of *The New York Times*, open to your participation however much or little you want. Each age group has its own comfort level and willingness to text message. Know where you stand. The groups are divided into three brackets that significantly differentiate one from the other. However, there's one tie that binds us all together: We date, therefore we text.

1. Late Teens

You're a Starbucks-sipping, aviator-wearing, Converse-sporting, sexually charged teenager with the latest cell phone. When we were your age we passed folded notes; you're doing the same but with text messages. Teens seem to be the most savvy when it comes to techno-relating because you started young. Texting is your world.

2. Mid-twenties

When you began dating, boys called to ask you out. You have since segued into dating in the flirtexting era. Therefore, you're a bit unsure how the rules have changed since texting took over the courtship. At first, you told everyone that you hated flirtexting and insisted that boys call. But then you realized that texting is *ah-mazing,* because it's a huge timesaver and an outlet for mad flirting. We think twenty-somethings actually have the best of both worlds because you have adopted texting as your new go-to for dating, but also know when it's appropriate to call.

In short, our book is for anyone who has a cell phone and wants to get it on. It is targeted toward girls, because— let's be honest—boys don't like to work on relationships! However, we know they'd be tempted to sneak a peek if this guide were sitting on their girlfriend's coffee table, so as a kind gesture to the men we text, we unveil our likes and dislikes through- out the book so that they know just how to text their way into our phones and, eventually, our hearts.

3. Early thirties and on ...

You thought texting was just a way to relay information, i.e., to tell someone you are on your way or to give direc- tions. You believed that e-mailing was the ticket that led to that phone conversation. But then you met a hottie-bo-body at a bar and he texted you out. And you thought, "I could have fun with this!" You are confident, self-sufficient, and know where you are heading in life. You know what you want and how to express it. Texting is the perfect tool for you to take what you know and work it through the flirtext.

an ode to *the rules*

In order to understand our approach, you must first understand where we are coming from. The two of us are "*Rules* Girls." If you aren't already aware of what being a *Rules* Girl means, it means playing a little hard to get, to get what you want. By not giving away too much at first, you leave him wanting more. *The Rules* teaches you that, through phone conversations and in person, the more mystery you maintain, the more interested he will be in you. We teach you how to obtain that same sought-after attention, via text.

With technology rapidly becoming the means by which we communi-date, we know that ultimately *The Rules* have to be adjusted to today's times. If we want to remain on top of our dating game, we need to keep up and adapt. We find it necessary to apply these old rules to modern times. For instance *The Rules* states: "Always end phone calls first." We agree with this rule and think it's essential to apply it to flirtexting conversations as well. For instance, if you're having a flirtexting convo with a current crush, it's best to let his text be the last sent. This also applies for starting up flirtexting conversation; you should allow him to start more of the conversations than you. It's about him pursuing you and not the other way around. You want his word to be the last, so he's left wanting more of you and your conversation. We consider this a texting cliffhanger—he'll want to fill his desire to communicate with you more.

We truly believe that *The Rules* will always be that by which we live and date. But today, we are dating in a flirtexting world. Keep the old rules, but learn and apply the new ones, so that you don't fall behind.

The Usual Suspects

With its non-threatening approach, texting has made flirting with lots of guys simultaneously more accessible. It's easier and quicker to weed through the many boys you have crushes on and see whether you want to take a relationship to the next level.

If you're single, this is who you are flirtexting ...

| **The Boy(s) of the Moment** | This is the boy you are currently dating or entertaining the thought of dating. Your flirtexts usually consist of funny, light jabs at one another. You're building a bond that'll lead to a series of dates and eventually, if you deem worthy, a potential relationship. You are very careful when composing your flirtexts to him and usually find it necessary to consult a friend. Totally normal and cute. |

Your "Guy Friend"	This is one of your close friends with whom you either hooked up with many moons ago or swear to everyone that you are "just friends" despite the fact that it's obvious there is sexual tension between the two of you. You're open and crazy flirting with each other because, let's face it, you're "good friends." Your flirtexts consist of playful teasing and mentions of how much you miss each other when you're not together. Take a tip from the theme in *When Harry Met Sally*: Men and women can't ever just be friends because the sex part always gets in the way.
A Guy You Briefly Dated	This relationship may have been as brief as a spring break hook-up, but there's no denying that sometimes it just feels good to flirtext with an old flame. Sure he may live far away now, or perhaps there just wasn't enough chemistry between you for it to work. In spite of that, it's always fun to have flirtatious banter with him when you get the urge. After all, he does make you laugh and his unwavering interest in you is very flattering. His admiration for you is always welcomed and you aren't afraid to hint around for it when you feel the need.

An Old Boyfriend	It's fun to flirt with him because, let's face it, he's hot and makes you laugh. As much as we like to look through old photo albums, we love to touch base with old boyfriends to relive those precious times shared together. Texting is the perfect way to keep in touch with old BFs, especially when you need a little boost of self-confidence. Your relationship with this old BF ended amicably, and even if he is involved in a new relationship you still keep in touch by flirtexting. There's an undeniable spark that you two will always share, and you like to see if it's still there from time to time. No, chemistry doesn't disappear even when one of you has moved on. Your flirtexts are commonly reflective of the type of things you said to one another when you were in a relationship.

Confidence:
You Can't Fake It

Confidence plays a major role in helping any girl land a great guy. It's the one characteristic that sets you apart from the rest of the pack and attracts men like magnets. There is something about a self-assured, intelligent, quick-witted girl who knows where she's heading in life that'll always turn heads. Confidence is a trait that you can't fake, no matter how hard you play "the game" or how good of a flirt you are. We have no doubt that with confidence on your side, you will always end up on top (pun intended).

Confident girls are sure of themselves inside and out. Everything about them exudes certainty and assurance. They are usually fearless and always go after what they want. They know their weaknesses just as well as they know their strong points. They're able to laugh at themselves and aren't afraid of putting themselves out there. They'd rather be up-front and take chances than not try at all.

Being a confident person shows in everything you do. It's in the way you walk, talk, and guess what, it's in the way you flirtext. Therefore, when you text, mean what you say. Stand by your ideals and values. Don't change your views for the sake of thinking someone else will appreciate them. Be proud of who you are, where you're from, and what you stand for. Having this awareness at all times, including when you flirtext, will benefit you tremendously.

You know how the saying goes: "When it rains, it pours." Ever wonder why it is that the second you're off the market, you become flooded with

> **Dating the Flirtexting Way:** Two people casually spending time together for the purpose of seeing if they want to eventually become exclusive. You are allowed to, and absolutely should, flirtext and date other people if you are not exclusive. The best thing you can do for yourself while single is to keep your options open. Once you have mutually decided to date each other exclusively (i.e., you only want to be flirtexting each other), then a talk will "seal the deal" for you to officially be boyfriend & girlfriend. (For rules on when flirtexting is considered cheating see Part Three.)

attention from eligible suitors? The aura of confidence around you, that you wear so well, attracts boys like honeybees to clover. The secure feeling that comes with being in a relationship is what helps you let your guard down and allows you to be your natural self. This makes you more approachable. The trick is exuding that same confidence in your flirtext.

Think Like a Boy

The two of us have been told many times that we think like boys, which we realize is an abnormality of the conventional chemical make-up. When it comes to dating, guys and girls typically have very different attitudes. Guys tend to take a more laid-back and carefree approach, while us girls sometimes complicate things more than we should.

Let's start with the boys. If a girl they like likes them back, they're thrilled. If she doesn't, they move on. Boys don't over analyze the things she said, nor do they drive themselves crazy trying to figure out why she turned them down. Guys roll with the punches in order to avoid drama, and for the most part, they aim to please. Easy breezy. This attitude is very apparent in their straight-to-the point flirtexts. Example: "when can I see you?" or "come over."

Now, girls, on the other hand, we're a bit more intricate with our approach. We tend to wear our hearts on our sleeves. We let our feelings get in the way of rational thinking, or *text when we know we shouldn't*. We lay our cards out too soon, displaying too much eagerness too soon within our texts. We are so used to getting our way that at times we have a hard time compromising, *if he doesn't text when he says he will, he's a goner.* We discovered that by thinking like a boy, we avoid unwanted stress and anxiety in our relationships. Displaying that same laissez-faire attitude within your flirtext will impress the guy on the receiving end and make you more desirable.

Guys love girls who are laid-back and easy-going. Having this attitude makes guys' lives easier and makes being with you more enjoyable. Try to remember this the next time a potential boyfriend (a PBF) flirtexts you some bad news. Say you got two fifth row tickets to a baseball game and invited your

> ### In any situation:
> *Being needy in your text = Turn off!*
>
> *Being chill in your text = Turn on!*

crush. You can't wait to impress him with your seats behind the catcher's mound and your newly acquired knowledge of A–Rod. A few hours before the game you get the following text from your date, "Hey im sorry something came up at work and I cant make the game tonight. i'll call you later." *Strike one.*

Here's where displaying a laid-back attitude in your flirtext will benefit you. Even though you're disappointed that he canceled, the trick is not to express your degree of disappointment in your response. Instead, think like a boy. Ask yourself how he would respond if the situation were reversed and you were the one canceling. Responding with, "No worries about the change-up. I won't hold it against your box score. I'll call in a replacement hitter to take the green light and batter up" will impress your crush and show that you are a cool girl who can roll with the punches. (We know we went over-board with the baseball terms but we wanted to give you options.) Or if you're a bit more forward, "No worries! Although I was looking forward to going

> *Bottom line, sporting a laid-back, easy-going 'tude will greatly benefit you. Having this mind-set will become apparent in your flirtexts and will help you land on the winning side.*

to first base with you. Next time;)" (insert first, second, or third base depending on how far you've gone). Watch as your nonchalant yet flirty response has him rescheduling the next time your team has a home game.

Bottom line, sporting a laid-back, easy-going 'tude will greatly benefit you. Having this mind-set will become apparent in your flirtexts and will help you land on the winning side. Give him a taste of his own medicine and watch as your bases become fully loaded. Now play ball!

Hook a Guy with Your Text: Show a Little Leg and Then Pull it Back

We all operate by the rule that we want what we can't have. It has proven to be true in life and business, but especially so in love. Therefore, it's no secret: *single guys want what they can't have*. Here's how they rationalize: If a girl they are interested in proves to be an easy catch (i.e., responds to all of his flirtexts immediately after receiving them and/or agrees to go on dates with him last minute) then they eventually get turned off by the girl's "too available" stat and move on. No need to analyze this because it's always the case. We say hooking a guy doesn't have to be a game, it's just common sense that can be taught.

Aware of men's short attention spans and competitive edge in dating, we've incorporated the "want what you can't have" style into our flirtexting. We've instituted this as a way to stand above the rest. By holding back your level of attraction to a guy at first, you set yourself up as a challenge, making him more intrigued by you. Then, in time, as you begin to open up and grow closer, your initial resistance will make him feel like he has earned something by winning your attention. Compare this to a young kid whom you let win a game of cards. They're so happy they won, they begin to like you even more.

To hook a guy in the beginning stages of dating, you need to give him a little taste of how amazing and fun you are but then play a little coy so that you set yourself as a challenge. We call this showing a little leg and then pulling it back. Entice him by making your first text exchanges with him light, witty, and flirty, so he doesn't feel intimidated. Once he begins to respond quickly, you have him hooked and now it's time to pull back.

For example, when a guy sends you a random text, one that he thinks is funny, clever, and certainly worthy of a response, don't respond. You aren't required to respond to every text. Pick and choose your flirtexts. He'll begin to wonder if your not responding was due to him "turning you off" by something he flirtexted. Boys have admitted to us they rationalize dead air the same way we do (only for not as long). He's wondering if you didn't get his joke or if his text didn't go through. He's rationalizing all sorts of crap in order to keep his ego intact. By not giving him the response that he wants or feels he deserves, he'll begin to want it- and you—that much more. If this has ever happened to you then you know exactly what this feels like. It's absolute torture until you get a response. We may play a few harmless games now and again, but we're not one to torture a man's soul, so when he texts again, respond away.

texting tip

Guys are very simple creatures when it comes down to it.

If he's not calling you, he doesn't like you. If he's not texting you, guess what, he's just not that into you. (We heart you Greg Behrendt!) We know this is hard to grasp, but at the end of the day it's reality. Don't read too much into things, what you see is what you get. Learn to take things for what they're worth.

If you think for some reason, "Well, what if he doesn't know I like him" and find the need to keep in constant touch with him as a means of assuring your undeniable like for him, then know you're blazing the wrong path, Sister. Boys, guys, men don't easily stray if they truly like a girl. Stay true to yourself and be a little patient. By not responding to his every text and taking your time to write back, he'll begin to want you more. Learn to be patient and try to resist the impulse to act right away. After all, good things come to those who wait.

texting tell-all

As a flirtexter, you learn not to tell a boy that you want to marry him and have all his babies from the get-go. You don't want to inflate his ego by letting him know how into him you are. Instead, you want to remain in the "I could take you or leave you because I have, like, a gazillion hot, amazing men in my life" category. In order to do this, we encourage playing a light game of hard to get. **Note:** If you are one of those girls who think that games are immature and pointless then pay extra special attention. When played responsibly, strategic games are actually very helpful and produce results. Girls who seem too eager and easy to please turn guys off. Know the right amount of game playing necessary to reel him in, and then pull back and be real.

Another easy way to show a little leg and pull it right on back is instead of responding to his flirtexting jokes with a typical "LOL" or "hahaha," hit him back with an equally witty (if not better) comeback. Guys are so used to girls feeding their egos by telling them how funny they are or by laughing at their lame jokes, even if they aren't funny, just to be polite. No ma'am, not us. Our money's on the fact that he's probably texted this "joke" or one like it to a different girl at a different time. Therefore responding with a typical "LOL u r so funny" is predictable and expected. Instead try responding in a way that will set you apart from the rest of the herd to keep him on his toes: "What's this nonsense about a chicken crossing the road?"

Teach Him How to Treat You

"It's pretty simple. If you write a girl late night and she writes you back, doesn't matter what she says, she wants to hook up."—**Jason**

Guys like rules. Really, they'd be thrilled to have a road map describing exactly what to do in order to get the girl that they want. You set the standard by treating others as you would like to be treated. In the early stages of courtship, it is therefore imperative that you set up expectations early enough so that he knows how to behave and communicate with you. This sets the tone for your relationship thereafter.

> **PBF (Potential Boyfriend):** **A crush. You could have known him for two years or two minutes.**

You can easily teach a potential boyfriend (PBF) how to treat you through text messaging. Say, for example, you went on one or two dates with a new crush. Then, one night around 1 A.M., you get the following late night booty text (LNBT), "hey...what are you doooin?" You have two options: You can respond within the hour and go meet up with him, or not respond at all. If you choose to respond, beware that you're setting yourself up to be placed in his "LNBT" category and most likely, you will forever be branded that way, no matter what "I don't usually do that" excuse you plead. Your second option is to not respond to his late night messaging. By doing so you're setting a tone of how you expect him to treat you. Your non-response will send a strong signal. He'll get the picture and call or text you during daylight hours. Don't feel bad about not responding. After all, it's 1 A.M. and you've just started dating. What does he expect?

In every and all relationships you form it is your responsibility to teach someone how to treat you.

And, may we add, what kind of girl does he think you are?!

All we ask is that you think about this next time a crush reaches out to you at a late night hour. Do you want to be the girl that he makes plans to see in advance and takes on fun dates, or the girl that he knows he can see late night without any notice or effort on his part? If you want the former, then set the standard from the beginning. If he likes you he will chill with the LNBT and contact you during non–booty text hours.

> **LNBT:** Late night booty text. When a boy texts you after 10 P.M. with a vague "whatcha doin" text, he's not making an effort ...

This is why early on you have to let boys know the behavior they can and cannot get away with.

texting tell-all

Volleying for Love

Our friend Jill was playing ping-pong with some friends one night. She spotted a certifiable PBF playing at the table next to hers so she decided to "accidentally" hit the ball onto his table. Whoopsey! The night ended with them exchanging numbers. The next day she received the following text from him: "Nice game Jill. Your backhand could use some work though. When do we rematch?" Score! Not only was she impressed by his pong game, but judging by his clever and charming text, she was sure he was a winner off the table too. Her flirtext response read a little something like this: "Actually, I let you win ... My trainer doesn't allow me to play with amateurs but I'll make an exception for you. Rematch next weekend."

Cut to one year later, and there were ping-pong tables set up at Jill and Matt's wedding reception. Matt finally found his match in Jill, on and off the court. Jill says it was their first exchange of awesome texts that set the tone for her and Matt's happily ever after.

YOUR SECOND FIRST IMPRESSION

Your initial text is one of the most important flirtexts you'll send. It's so important in fact that we refer to it as "your second first impression." You can tell a lot about a person by what they text. The only way for him to get to know you at this early stage is through your texts. Therefore, the better you text the more he will like you. It's as simple as that. In Jill and Matt's case, they were able to reconfirm their mutual attraction and similar sense of humor right away through their initial flirtexts.

Your initial response text is a chance to show your degree of interest and to showcase your personality. If his initial text to you hits it out of the park then you want to respond with equal or greater enthusiasm and awesomeness.

Your initial response text is a chance to flash your playing cards so he knows what he's up against.

Here's the point of your initial response text:

1. To confirm your interest in him.

2. To make him smile, laugh, and blush. Do all three and you're a sure thing.

3. To reflect your personality: funny, flirty, sarcastic, etc.

4. To try to out-do his initial text. You may not always hit the mark, but at least shoot for the stars. If his initial text was boring, then you have little work cut out for you.

5. To get a date invite.

When a boy nails a great text, you want to surprise him by throwing something back that is even better and

> *You can tell a lot about a person by what they text.*

has you coming out on top. By outdoing his initial text, he becomes pleasantly surprised by your effort and will want to come back for more.

Here's what to look for in an initial flirtext from your crush:

1. Expect an initial flirtext within 24–48 hours of giving a guy your number.

2. He will attempt to re-establish the connection that you made in person. He'll do this by bringing up a conversation or something that happened when you met.

3. He will try to feel out what your schedule is like to know when you're free to go out.

PART TWO:

game on

Creating the Best Possible Text (BPT)

"You're in control when you text. You know exactly what to say and how to say it, and you can take your time if need be. That's an advantage, especially for those not very quick with the tongue (I happen to be a master with mine—hee hee)."—**Nat**

putting your best text forward

There are some things in life that should be avoided and awkward phone conversations are one of them. For everything else, there's texting. Most boys feel talking over the phone puts them on the spot to constantly entertain with funny jokes and tantalizing stories. Take this example:

Jenna: Hello?
Ben: Hey Jenna, this is Ben.
Jenna: Who?
Ben: Ben. We met the other night at the bar.

Jenna: Oh yeah, hey. What's going on?

Ben: Oh, not much. So, how's your foot?

Jenna: Excuse me?

Ben: Remember? You got stepped on while we were dancing. We almost got into a fight with the girl who did it.

Jenna: Oh yeah, (awkward laugh) I forgot about that. Yeah, it's fine. A little bruised but I'll live.

Ben: Oh good. Thought they might have to amputate. (dead air)

Jenna: (another forced laugh)

Ben: Uh, just wanted to call to say I, umm, I wanted to see what you're up to later tonight or this week. I was, uh, thinking we, you know could maybe catch a flick or grab some dinner … (long pause) if you were free ya know? (clearing throat)

Jenna: Oh cool, yeah, let me look at my calendar and call you back.

Later … no call back

If you ask a boy how he feels about phone calls, we'll bet you he says he *hates* talking on the phone. Boys *love* to text because they feel much less pressure than when calling. Say goodbye to blurting out dumb phrases and pointless

> One of the biggest advantages texting has to offer is complete control over what you say.

Recipe for Success

Coming up with the absolute perfect text and sending it at just the right moment is our secret to winning your guy over flirtext.

stories all for the sake of avoiding awkward silence. For all these reasons, boys love to text, so girls MUST learn how to flirtext.

We created a guaranteed way to land your man through text. It's our recipe for love, if you will. The recipe contains two main ingredients: creating the Best Possible Text and knowing when to press send. We can be impulsive when we like someone. This can translate into our texts. The key is making sure your text is well thought through and not delivered on impulse. A great text draws attraction and more interest toward you. It makes you stand out, look smart, and shows that you are with it. A girl who can nail a great text is one that any guy will want to text more.

Following is what the same awkward phone convo between Ben and Jenna would sound like via flirtext:

Ben: Jenna, you've got dance moves that would put Paula Abdul out of work. How's that crushed foot of yours doing today?

Jenna: Hey! I did outdo Paula, huh? The foot's a little bruised but nothing some ice and a cold drink wont fix!

Ben: How about I'll get you that drink if you promise not to dance too close to me and make me feel bad about my dancing skills?

Jenna: Deal. But I can't make any promises about the dancing. You saw me last night—I'm a machine;)

The first part of successful flirtexting begins with creating the Best Possible Text (BPT).

> BPT: **A text that says exactly what you want to say and in the best possible way. In order to create this, your message has to be strategically thought out and aimed at getting your desired response.**

You can tell a LOT about a person by what they text, making it a necessity to always put your best text forward. As quick as we are to put our best foot forward when meeting someone we like, we must remember to do the same with our texts.

Putting the right ingredients into your flirtext will show him that you are interested, as well as keep his interest in you piqued. Luckily, text allows you the time to compose this perfect message!

Look Before You Leap

Texting is instant and as a result encourages impulsive responses. When you're sending a text to a boy, taking two seconds to consciously think about what you are saying, instead of just typing the first thing that comes to mind, can help ensure your text is the best it can possibly be.

texting tip

A valuable lesson we've learned in life is to try and take a step back to think about what we're doing and what it is that we want at any given moment. Taking a second to consciously think about our actions before doing them can help reduce mistakes and curb our impulsive nature. For example, if you are at dinner and the waiter sets a huge plate of french fries in front of you, taking a second to recognize if you actually want a fry before impulsively reaching for one can make all the difference. The same goes for texting.

You always want your flirtext to include just the right amount of humor, mystery, and sass to make you seem desirable and to leave your crush wanting more.

how to create a best possible text (BPT)

1. *Decide what you want:*

   (As if there were any other option for number one?!) Decide what it is that you want to get out of this flirtexting conversation. Do you want him to ask you out? Come over? Go to drinks? After you decide what it is that you want, writing a strategic message will help you get it.

2. *Forecast obstacles:*

   Think about anything that could get in your way of getting what you desire. Try to remember if he said anything about his weekend plans when you spoke to him earlier. Did he say he was going out of town? Is it his best friend's birthday? It's always a good idea to have a little 411 on your honey's weekly activities before you hit him up. You wouldn't want to ask him to do something when you can mathematically prove he will turn you down. This step will ensure your attempt won't be wasted.

3. *Brainstorm BPT:*

Now that you know what you want, and can't predict any obstacles that would get in your way, create a BPT and get it! In order to do this, you must never write the first thing that comes to mind. If you draw a blank or have doubts about what to write, ask a friend. If that doesn't help, check the Internet for inspiration. Do whatever it is that you need to do in order to feel like you have narrowed it down to the absolute BPT.

- **Ask a friend:** *It's always a good idea to get a second opinion when concocting a BPT.* It's difficult being flirty, sassy, fun, and impeccably dressed 100% of the time. In times when you're not on top of your game and need a clever flirtext, friends are your best resource. They can help turn your run-of-the-mill text into a BPT. Since they are not emotionally involved in the relationship like you are, their ideas are usually more playful and less needy. Best part: you get all the credit! Take into consideration that what you may think is a sweet, lighthearted message may read as "trying too hard" to someone else. Your BFF will help you maneuver this situation—after all, that's what they're there for. Use them.

- **Surf the web**: *When in doubt, Google.* If you're having difficulty thinking of a BPT, and your BFF is at yoga, we recommend the Internet for inspiration. For example, try:

  www.thedailyshow.com—Current events and pop culture news with a comedic twist brought to you by Jon Stewart.

  www.someecards.com—E-card Web site with brilliant one-liners, such as "I'm somewhat skeptical you're laughing out loud as much as you claim."

  www.imdb.com—Movie Web site where you can pull quotes from any movie or TV show *or* reference our Movie Quote section in the back of this book.

  www.perezhilton.com—Latest celebrity news brought with humor. You can grab some of Perez's lines and make them your own. He won't mind.

4. *Leave an option out:*
   A *Rules* Girl always leaves her options open. In cases when you're asking to make plans with a boy, if possible, try and create an "option out" in your text in case he comes up with something better or is unable to make it. Meaning, don't ever commit to something before he agrees to it

first. For example, never text, "I'm going to the movies on Friday night do you want to go?" That statement is too definite. If what you want is to see him, movie or not, then instead text, "Thinking about catching a flick on Friday. Interested?" This leaves your plans more open-ended, giving him the opportunity to suggest something better.

Know when to press send.

BPT thought process

You decide you want to see a movie with him tonight. You can't recall anything important he'd be doing since you talked to him yesterday, and he didn't mention anything. You know he's a Johnny Depp fan and it just so happens Depp's new movie is debuting this weekend. You concoct a text that's enticing enough to ensure a "yes" or at the very least a date in the near future.

It reads: "Any interest in seeing J.Depps new flick? Popcorn on you, Gummi Bears on me."

knowing when to press send

"Because it helps me distinguish the desperate from the not desperate girls. Response time says a lot about a person."—**Josh**

You just received a flirtext from your crush. You're stoked because you've been thinking about him all day and were wondering if he was doing the same. The minute you receive his text you begin writing the first response that comes to mind ... and then press send. Five, ten, thirty minutes go by and he still hasn't written back. You begin to F-L-I-P out and are certain his silence is due in part to your lame, impulsive text. You can't believe you wrote back, "Stop it! That's soo hilarious Pete!"

In the quiet meantime, you start driving yourself crazy coming up with a half dozen better flirtexts which you are certain would have elicited a faster response. You swear to yourself that when *and if* he finally writes back, you'll be more patient with your next response to ensure the LOL you so rightfully deserve. Or at the very least, a timely response.

Now that you've gotten your message in place, the second part of successful flirtexting is knowing when to press send. Unfortunately, part of being a girl means getting overly excited about a simple little text from a cute potential boyfriend. A text can escalate that already neurotic butterflies-in-stomach feeling you have for a boy, and can drive the common sense right out of a girl. In other words, upon receiving his text, you feel the need to respond instantly, 'cause let's

> *Don't let the girly-girl in you get in the way of your game and control. If he texts you, then the ball is in your court and you hold the power. We recommend you try and hold onto that power for as long as your manicured hands can.*

face it, you l-o-v-e him. You respond with the first thing that comes to mind and send it. We call this a Bad Idea.

> **Bad Idea: Immediately answering any flirtext from a crush.**

Just because it's speedy technology does not mean you need to be instant with your response! It's important for you to realize that the first text you think of is in all probability impulsive and driven by emotion. The key is to tone down any excitement you may have and resist the urge to write the first thing that pops into your head. There certainly is no rush when it comes to flirtexting.

Unlike phone conversations or F2F interaction, where you are put on the spot to be clever and witty, text gives you the power of time. The secret here lies in the way you use this time. By having the element of time on your side in flirtexting, you hold the power and upper hand. Yes, it's as simple as that.

Admit it. In the back of every flirtexter's mind there's an unspoken timeline pertaining to when one texts and when one responds. It's all really a game of catch-me-if-you-can. Responding at just the right moment is *key* when trying to land a man via text. If you respond too soon you could come off as looking desperate. Wait too long and you might come off uninterested. Knowing *when* to press send is just as important as what you are sending . . . if not more so. Allow us to explain.

While the phone call/F2F conversation is more instant, text allows you to determine your speed according to your desire. You can respond back to text messages *instantly,* exuding interest and eagerness; *delay responses,* which conveys a sense of mystery and that you're hard-to-get; or ultimately *not respond at all,* which can mean either of two things: you're not interested in him *or* you're not interested in what he texted you, be it simply boring or sloppily in 3 A.M. drunk mode. By using time in this way, you adjust your game to your needs, whether you want to be straight and to the point

> *Text gives you the advantage of time and, therefore, the power.*

texting tip

Stop!

Do not respond immediately to any text. Even if you think you have the perfect response, chill. Call your BFF, watch some Tivo, take a run ... whatever! Do anything to distract yourself. Remember, he's lucky to be getting a flirtext from you. Make him realize it by making him wait. Leave him always wanting more.

or a coquette. Ball is in your court; use it to your advantage.

In essence, the longer you can hold out to respond, the more he will like you. The same way *The Rules* told us to wait a day or two before returning his phone call, we're telling you to keep this same state of mind—hold off a bit—when responding to his text. This signals that you are not an easy catch ... and we can't argue the fact that boys love a chase.

Patience is an undervalued virtue that'll help any girl, any time. How many times have you sent a text because you felt the need to respond right away, and fifteen minutes later wished you had written something else? Trust

Waiting it out means he's sweating it out.

your gut. Wait until something brilliant comes to you even if it means responding the following day.

> **Waiting It Out: We're not talking waiting a week to respond. Do this and he'll think you're not interested. Every flirtexting situation differs, in which case the response time will vary. However, the longest you should hold out when responding back to his flirtext is twenty-four hours. Anything beyond that reads as uninterested.**

Here's why delaying your response works to your advantage. When you hold off on your response, your potential boyfriend may start second guessing his text, as well as his approach. While you're composing something brilliant, he's going crazy wondering if you still like him. Boys have revealed to us that they go nuts waiting for our response to flirty texts just like we do. He may not admit it or show any signs of it, but he's freaking out a little bit. Guys are a lot better about keeping their emotions under wraps, while girls find the need to talk it out. They hide it, we analyze it. We're different creatures. If he admitted that it drove him crazy waiting for your response then he would lose a little game and a whole lotta manhood. So when you come back with the perfect response, he'll be quietly thrilled! Trust us, by not responding instantly you'll reap the benefits in the long run.

The delayed text works best in the most critical of courting situations. Whether you are head-over-heels or just interested in getting to know him better, the best thing you can do for yourself is hold out on responding right away. The saying goes as follows: "We all want what we can't have." Take this and turn it to your advantage.

We KNOW holding out on responding is so hard to do. But may we remind you, it's not like he showed up on your doorstep with a half dozen roses and a cupcake. It's a text.

Yes, texting back immediately leaves you with the instant gratification that you so desperately need to feel. Remember, this feeling is followed shortly by a slow fall into depression when you don't hear back right away. It's like a sugar high you don't want to be on. What you want to do is to feed his ego to make him feel like he's the luckiest guy in the world when you finally write back.

This technique is probably similar to the strategy you used to get his interest the first time you met. You most likely flirted with someone else in his direct eyesight to make him notice you. Finally, after making him sweat the whole night, you give *him* your number and not the other guy. This gives him the ultimate satisfaction because in a sense

> *Quick reminder: Jane is that desperate girl you don't want to be!*

he'll feel like he's won. Not that it was a game or anything ;).

Here we list different timelines for texting scenarios and what it means by how long we take to respond. These approximate timelines pertain to budding relationships. The texting timeline does not apply if you are texting back a friend or family member. And it doesn't count if you're in a serious relationship. We advise using them in the initial six weeks of dating. Knowing when to press send is key when trying to hook-line-and-sink your man.

TEXTING TIMELINES:

when to respond

1. His text—"What are u doing tonight"—at 5:30 p.m.

	When you respond	What it means about you
Jane.	5:32 P.M.	You love him and now he knows it and has all the power. Don't do this.

The Romantic.	**6:00 P.M.**	You like him a lot. You want to hang out with him and want him to know it.
The Chill Girl.	**7:30 P.M.**	You think you like him but don't really know yet. Perhaps you'll know more based on his response.
The Bad Ass.	**8:30 P.M.**	You like him but want to field your options. He's not the only one to ask you out tonight. He better have a good plan to keep your interest.
The Shameless.	**1:30 A.M.– 5 A.M.**	Oh, sweetie (we still love you).
The Busy Girl.	**3:00 P.M. the next day**	By taking this long to respond you probably had a lot going on. You're giving off a catch-me-if-you-can 'tude.

2. His text—"Dinner tonight?"—at 3 P.M.

	When you respond	What it means about you
Jane.	3:05 P.M.	You're eager to see him and made it quite known. It's mid-day and you are a working girl ... so act like one and don't respond right away, even if you do want that dinner. Mr. Last Minute isn't the only meal ticket in town, let him know it.
The Romantic.	3:30 P.M.	You're excited he asked you to dinner and are already planning what to wear. You hope he picks a romantic spot.
The Chill Girl.	6:00 P.M.	You just got off of work and you're considering meeting him later. You text him back and ask what time he's considering doing dinner.

The Bad Ass.	**8:00 P.M.**	It is very clear by now that you're not going to dinner with him. By responding this late on, you let him know that you're not one to succumb to last minute dinner invites. You should, however, apologize for getting back to him so late and say that you'll be out at X tonight if he wants to come by after dinner. Watch him run over and meet up.
The Busy Girl.	**Noon the following day**	He asked you to dinner the day of and you already had plans. We know you were busy but he did invite you to dinner. It's not polite to leave him hanging, even if it was last minute.

texting tell-all

Plan B?

If a boy texts you to ask about dinner a few hours before the reservation, keep in mind his plans may have fallen through which means you're plan B for the evening. When a boy likes you and wants to take you out to dinner, he usually asks you a few days in advance.

3. *His text—"What's going on?"—2 P.M. after two weeks of silence*

	When you respond	What it means about you
Jane.	2:15 P.M.	Who are we kidding? You're so happy he texted and make your enthusiasm well known in your response time.
The Romantic.	3:00 P.M.	You kept a positive outlook and promised yourself if it's meant to be, he'll re-contact. You write back in hopes of seeing him very soon.

The Chill Girl.	**5:00 P.M.**	You're psyched that he's back in the picture, but you're taking your time with letting him know that.
The Badass.	**The following day**	You give him a little taste of his own medicine. And don't make it easy for him to get ahold of you quickly.
The Busy Girl.	**Won't respond till his second attempt at contact**	He's been out of contact for two weeks. WTF. Obvy he's interested in you because he's re-establishing contact. But just because he is, doesn't mean you're gonna jump. Therefore, you wait until he texts again to earn your attention a bit more.

texting tell-all

Booty-Text

If a boy texts you to see what you're up to any time past 10 P.M. he only wants to make out with you late night. If you like this guy and feel it's too early for this sort of causal meet and greet then don't respond back until the following day. This will train him to know that you are just not that kinda girl. If you do respond that's okay too. Just know that he will most likely only think of you when it's late at night and ... well, you know what we're saying.

the waiting game

"I love the anticipation of response. You know the feeling you get after you send a text and start wondering what the recipient is thinking ... I always know that I really like a girl if I'm stressing over my text after I pressed send."**—Pete**

It's Saturday afternoon. You and your PBF were texting back and forth earlier that day about insignificant nothings. Now it's 3 P.M. and you want to know if you are going to see him tonight. After all, you're not about to waste a new going-out top if he's not going to see it. You decide to text him to see what he's up to later, so you write, "Big plans tonight?"

Ideally he will respond within thirty minutes with "you tell me;)." But he doesn't, so you wait. And wait. You go to the gym for a spin class and hope no one sees that you set your phone in the cup holder. Lame. Class is over and still nothing. You head to Starbucks for some caffeine, because at this rate it's looking like a long night. While enjoying your coffee your phone buzzes and as you lunge into your bag you burn your hand. UGH! It's only your BFF asking what you're wearing tonight. Finally you hop in the shower and decide that if he doesn't write back by the time you're out, it's his loss and you're moving on. Then despite him, you decide you're going to wear the new top anyways because you're on the prowl to replace "Mr.–I'm–too–busy–to–text–back." Whatever.

It's Just a Text

Here's the thing: Boys in general are naturally more nonchalant about responding in a timely manner than girls are. When he's at the gym, he's at the gym. When he's eating lunch, he's eating lunch.

He might read your text immediately, but he won't respond until he believes it is appropriate for him to do so (i.e., after the gym or after lunch). Boys don't get giddy when they open a text, read it to everyone around them, and ask their spotter at the gym for help on creating the best possible text. Can you imagine?

This time lapse may appear to you like he's playing a game or not that into you, when in reality it's just him being a guy. Train yourself to think differently about why he hasn't responded. Know that his non-response has nothing to do with you and everything to do with him (unless, of course, you sent a lame text). Whatever you do, please don't send him another text just for the sake of getting him to respond. Boys deem that as being desperate and it's a huge turn off. Put it this way: Jane would send another text and then one to his friend trying to locate the guy. Don't be Jane.

Try not to let the time lapse get to you. Because let's face it, we know what happens to you when it does. He'll have you second guessing your last text to him and re-reading it a million times trying to figure out if his delayed response is due to your last text. Thoughts begin racing of: *Was it something I said? Was I rude? Did he not get my joke? Did his battery die??* Not only are these frantic thoughts a huge waste of time, but they are all

Did you get my text?

Don't believe boys' excuses, such as "I didn't check my messages until now," or "Sorry my phone died—I just got this message." We've all used these and know they're lies. He got your message and just didn't respond. Be wary of boys who make these excuses. Therefore, refrain from asking "did you get my text?" because it's a lost cause.

texting tip

pathetic excuses we come up with to try and help ourselves understand the reason behind his tardy response. You should understand that the tardiness is meaningless and you should stop worrying.

The act of texting is casual. Therefore, taking a little time to respond back is permitted and even admired. Try to remember this during the time you are waiting for his response. Even if you're secretly going crazy inside, the key is to create a laid-back attitude about the situation, much like our male counterparts do. Set the phone aside and go on with your day/life. Show that you are a cool girl, who has better things to do than wait around in agony for a text response. We guarantee that he will respond when he can. Just be patient.

Realizing you hold a lot of power in that little cell of yours is key to a long road of successful flirtexting.

A well-worded message sent at the time you desire can help navigate the direction where you want your relationship to go. Remember, what you send is just as important as when you send it. The time you take in responding says a lot about your character. Alert: Your popularity may improve by using this formula. Don't be surprised if your inbox is full. . . . and no, we won't pay your next phone bill.

FLIRTEXTING TIMELINE:

from digits to dinner

saturday	You meet a hot new crush and he gets your digits.
sunday	You check your phones a few times to see if he texted, but he didn't.
monday around noon	Crush sends you an initial text that is encompassing: witty, charming, clever, and flirtatious enough to make you fall in love.

monday night	Crush calls or texts to ask you out for Thursday night, to which you respond with a "love to."
tuesday	Nothing.
wednesday	He calls you to confirm your date and affirm his interest.
thursday	You exchange flirtexts throughout the day, which gets you excited about seeing him later that night.
friday	You text him that you had a great time last night. Then he writes back something funny that happened when you were together.
saturday	See last Saturday!

PRE-PACKAGED TEXTS:

clever responses to his common texts

Since we know you are busy ladies, we took the liberty of listing some BPTs to boys' most common flirtexts. Like us, they are mood dependent. The content of our text will depend on many things: the time he sent his text, how sincere it is, and whether or not we are having a good hair day.

These pre-approved and ready–to–use flirtexts are sure to get the response you want and then some;). And, by the way, feel free to take all the credit for them. It's what best friends are for.

Added bonus: We asked all of our hot guy friends to tell us what they *really* mean when they text these things. We have listed their candid and truthful answers, verbatim, under "what it means."

1. His text: "Hi"—and nothing else.

What it means: He's basically saying,"I'm here, don't forget about me…"

If you're feeling	Respond like this
Flirty:	"Hey yourself;)" *or* "Hey handsome"
Sassy:	"You're gonna have to do better than that" (calling him out on his lame attempt)
LOL:	"Who is this? Uncle Leo?"(Classic *Seinfeld* quote. Boys love *Seinfeld*.)
Straight to the point:	"Hey u what's happening?"
Jane:	"Heyyyy"(A pointless response to his pointless text … and mind your y's).

2. His text: "What are you up to tonight?"

What it means: He's beating around the bush and wants to ask you out. He's hoping you are free and will wait to see what you're doing before asking.

If you're feeling	Respond like this
Flirty:	"Not sure. What'd you have in mind Mr. Party Planner?" (Ball's back in his court … where it should be!)
Sassy:	"Whatever it is, it won't include boys who text me last minute asking what I'm up to tonight;)" (Slacker!) *or* "Ridiculously busy. Need to update my Facebook profile. You?"
LOL:	"I don't know. Was thinking about going to Home Depot. Buy some wallpaper, some flooring. Then maybe Bed, Bath, & Beyond. U?" (Tip: When in doubt movie quotes are always a good choice) *or* "Just finishing up our scrapbook" (HA! Obvy you aren't making a scrapbook of the two of you. You've been on two dates!)
Straight to the point:	"You tell me." (He's blushing)
Jane:	"No plans" (Social suicide, ladies!! Always have plans even if you are sitting at home watching Lifetime with zit cream on. He doesn't know that!)

3. His text: "Do u want to see a movie?"

What it means: He's into you and wants to take you on a date but doesn't feel like being on the spot for a two-hour dinner. Taking you to a movie is less pressure.

If you're feeling	Respond like this
Flirty:	"Only if you'll hold my hand during the really scary parts!"
Sassy:	"No thanks I've seen one!"
LOL:	"Yes as long as it's either —— or —— Because if you choose anything else I'll fall asleep. So excited to see what you choose!"
Straight to the point:	"Sure. You pick;)"
Jane:	"Absolutely. What movie and when? Are you picking me up?" (Don't be too eager and bombard him with questions. He asked you out; assume he has a plan.)

4. When you want to see him that night and he's not asking you out!

What it means: He's either really busy or not that into you. (Fingers crossed for really busy.) Don't fret, he could be having a moment of insecurity/loneliness/vulnerability and will text you in the near future. However, this isn't always the case. You should begin to explore other options.

If you're feeling	Respond like this
Flirty:	"I know a really cute [insert your hair color here] who will be at —— tonight. U should stop by and check her out."
Sassy:	"—— tonight. Get involved."
LOL:	"People I want to see tonight, for 400 please. Answer: Who is [insert full name of recipient]?"
Straight to the point:	"Buy you a drink tonight?" (No guy in his right mind would turn this down.) *or* "So I had your secretary clear your calendar and MapQuest you directions to the party I'm going to tonight. See you there!"

| *Jane:* | "Hey!! What are you doing later?" (Lame. Too generic. Boring. Need we say more?) |

5. When you've been texting back and forth all day and suddenly he stops texting, leaving you hanging ... Text one of these to get a quick response.

What it means: See Scenario 4.

If you're feeling	Respond like this
Flirty:	"I'm not sure I like your attitude."
Sassy:	"Sorry, did you not hear me? Should I type louder?" *or* "Should I assume you ran out of text minutes?"
LOL:	"Are we playing the quiet game? Because if we are, you're really good at it!"
Straight to the point:	Re-send your last text and act like your phone is messed up.
Jane:	"Are you ignoring me?" (Again, a case of social suicide.)

6. *His text: When he name drops and instantly turns you off. "Hey. Going to —— tonight. [Insert name of trendy celeb here] is going to be there. You should come."*

What it means: HE'S A TOOL. He thinks you're a cool chick and feels the need to impress you in order to compensate for his low self-esteem. Or, he thinks you are superficial and can pick off the low-hanging fruit with some materialistic bait.

If you're feeling	Respond like this
Flirty:	"Wow I didn't realize who I was dealing with … you're kinda a big deal huh?"
Sassy:	"Who's that?" (This will put him in his place)
LOL:	"Hang on I think you dropped something. Oh never mind, it was just those names."
Straight to the point:	IGNORE his lame attempt to woo you through who he knows. Hopefully he'll get the hint when you don't respond and try a second attempt, with a lot more class.
Jane:	"I'd love to meet _____ sometime. I'm such a huge fan!"

7. *His text: How to handle a cheesy/mushy text:
"You are so cute. I miss you when you're not
around."*

What it means: He likes you but be wary. He may
just be lame and cheesy.

If you're feeling	Respond like this
Flirty:	"Who told you that cheesy lines are my weakness?" *or* "Are you reciting lines from *Casablanca* again? (Throw an equally—or more—cheesy line right back to poke fun a bit and see how he reacts.)
Sassy:	"I know, aren't I adorable?" (Boys love a little cockiness. It makes them laugh and heart you more for your confidence too.)
LOL:	"Yeah, I'd miss me if I wasn't around too."
Straight to the point:	"That's so cheesy. And I like it!"
Jane:	"Aww you're so sweet, I miss you tooooo☺" (Excuse us while we throw up!)

8. *His text:* When you got into a little argument and he texts "Are you still mad at me?" or "Is everything cool with us?"

What it means: He f'd up and knows you are pissed off. Either you are ignoring him or he is afraid that if he calls, you will blow up again so he's texting to test the waters first. Depending on your response, he'll know how much he can get away with in the future.

If you're feeling	Respond like this
Flirty:	"Ask me again when you show up on my door step with fro-yo and *The Notebook.*" (This is fine in instances when you are OVER being mad.)
Sassy:	"Who is this?" (Obviously you're still mad.) *or* "Is Al Gore GREEN?" (Any updated pop culture news will work here.)
LOL:	"Ask me again in 5 minutes." (Keep doing this until he gets annoyed and calls. Something he should have done in the first place.)
Straight to the point:	Don't respond. He will get the hint.

Jane:	"No. I could never stay mad at you! <3" (If you say this, he'll do whatever he did to piss you off again. Guaranteed.) *or* "Yeah im mad. It was really disrespectful and I cant believe u would do that to me." (Don't have "feelings" talks over text.)

9. *His text: When he's been flirtexting you off the chain and you want him to call.*

What it means: This may be a problem. Or, he may think you prefer to communicate non-verbally. Drop a hint that you are not one of those girls and see if he responds in kind.

If you're feeling	Respond like this
Flirty:	"I think we're ready for the next step. I'll wait by the phone and hold my breath. Ready, go!" *or* "I'm doing something with my hands. Call me."
Sassy:	"FYI—Unlike Verizon I do not offer unlimited text plans" *or* "You know these things also place calls? We should really be adventurous with our phones."

LOL:	"All this texting is giving me carpal tunnel. [Insert digits here]." *or* "T-t-t-today Junior: [Insert digits here]." (*Billy Madison* movie quote)
Straight to the point:	"I'm running to the gym [or out to dinner]. Call me later." (Being the first one to end the conversation is the best way to leave him wanting more. If he likes you, he'll follow up on that call later.)
Jane:	"I really wish you would call me instead of texting." *or* "Hey do u think u can call me later?" (Don't put him on the spot. This sounds needy and eager. Both turnoffs.)

10. His text: "When do I get to see you again?"

What it means: He wants to see you.

If you're feeling	Respond like this
Flirty:	"I don't know. You tell me."
Sassy:	"Depends. Were you thinking coral diving in Malaysia or more along the lines of dinner and a movie?"
LOL:	"When Britney Spears goes bald. Oh ... shit. I'll be ready at 8." (Any improbable pop culture related news feed will work here.)
Straight to the point:	"When you make reservations."
Jane:	"How's tonight?"

Flirtextiquette

"It's fun making a girl blush just by texting a few simple, short things that you know will turn her on from past experiences you have had with her."—Justin

If you're flirting with the idea of flirtexting or are already a serial flirtexter, we urge you to mind your *p*s and *q*s and read up on the dos and don'ts of flirtexting. Practicing proper text etiquette, or *flirtextiquette*, will ensure a successful run of flirtexts. Flirtextiquette is a set of manners that will impress when used. We urge you to read up, manner up, and then text up.

When you should text:

1. *At the beginning of a relationship* to get to know him through flirty, witty banter.

2. *Light date* invites for a movie/drinks.

3. *Changing date* arrangements;
   i.e., time/place or when you're running late.

4. *Post-date* courtesy text.

5. *When you're thinking* about him and want
   him to know it.

6. *When you want to make him laugh.*

7. *When you want to catch up* with an old
   fling to see how he's doing but don't want to
   have a conversation over the phone.

8. *When he's calling* you and you don't like
   him, but feel the need to respond by making
   little effort.

9. *When you want to tell him* something
   that you're too shy to say in person or over the
   phone.

10. *When you're long distance and
    texting* makes you feel closer while keeping
    your minutes down.

When you shouldn't text:

1. *While on a date, in front of your date.*
   You may check your messages and quickly respond while you are on a bathroom break or if he gets up from the table.

2. *When it's really important.*
   Certain things should be said in person. Like: I want to break up, I love you, Will you marry me, I'm pregnant, etc. It's insensitive and shows a lack of respect to belittle such important topics to a text message.

3. *When you're angry.*
   Text is not the time or place to start a heated conversation or to talk about your current work/family/pet status. If you must, you may send a pre phone call text like: "Hey. Call me when you can. Need to chat …" This lets him know something's up.

4. *When it's 3 A.M. and you're drunk.*

5. *RIGHT after fooling around.*
   Pillow-talk time is not the time to be checking your cell. It shows lack of interest and will make anyone lying next to you begin to feel self-conscious.

rule for canceling a date over text

Rule of thumb is to cancel the same way he asked you out. If he called to ask you out, then the right thing to do is to call back to cancel. If you send a text instead, then you're giving him signs that you are not that into him.

If he texted to ask you out then it's absolutely appropriate to text to cancel. Just keep in mind that date arrangements made over text are more likely to be canceled than if they were made by phone. Like we said before, we are held less responsible for what we say and for plans we make via text. They are just as easily canceled as they were made.

post–date courtesy text

Back in the day, the Boleyn sisters would hand-write King Henry thank you notes for hosting them at his castle. Today, they'd text him to ensure a future invite and position in his court. Any time a boy pays for your outing experience and you're still

Rain Check Requests = Not that Interested

It's the oldest trick in the book. We agree to a date at first because we have nothing else to do on Wednesday night. But when Wednesday afternoon rolls around, and we decide we'd rather take a nap than get drinks with Matt.

So we send a text saying that we are terribly sorry, but something came up (i.e., you're feeling sick, your friend just broke up with her BF, your cat ODed on your Ambien, whatever) and we ask if we can have a "rain check." HA! The truth is, if we really liked Matt then we would have been planning our outfit the night before and getting a manicure the day of.

Attention, ladies!! The same applies when he texts asking you for a "rain check." Our money is on the fact that that check will never be cashed. It's a polite way of saying "I'm just not that into you." If he was interested and needed to reschedule he would have called. A text in this case is a cop-out.

So if he ever texts you, "rain check," take note and begin exploring your other options. It's not you, honey, it's him.

in the courting stages, it's common courtesy for you to text a thank you post-date. The post-date courtesy text is thoughtful and should be quick and easy, not to mention the least us girls can do. He did just pay for your all-you-can-eat meal and that made-to-order mojito you had the bartender whip you up. A simple "thanks for such a yummy dinner had a great time stud!" will reinforce that you're a polite girl (whom he can bring home to Mom) and give him the go-ahead to ask you out again.

> **Marathon Texting:** Non-stop, all day flirtexting with a crush. Just make sure to give him time to sweat between texts!

more texting no-nos

Non-stop text messaging.

While times exist where non-stop texting with a boy generates butterflies and excitement through much of your lackluster day, don't expect the following day to deliver the same. Boys tend to get burnt out from texting quicker than girls do. If a guy has a hectic morning of e-mails and phone calls,

Degree(s) of interest

If you just had the best date of your life—we're talking can't-stop-smiling-you-lost-five-pounds from-all-the-excitement date—then it's BEST to give him a call the next day to say thank you. If you really had that much fun, then we're sure he did too and would appreciate the call/confirmation. However, if your date was nice but nothing to write home about, then a sweet thank you text will suffice the following day. If the date is a complete flop and you hope he moves to Istanbul next week, your manners still shouldn't get in the way of your feelings. Texting "Seth, thank you again for dinner. Speak soon." the following day will let him know where he stands.

texting tip

he'll tend to put the phone away for the afternoon to recoup, even if he really likes you.

You can generally sense a person's interest in a flirtexting marathon. For instance, if you text a boy first and he merely writes back to your text with a brief, non-flirty response when he normally gives you more, know to stop texting him and wait for him to respond later. Commonly, a boy who's interested in marathon texting will most likely text you first and throw out something extremely flirty and well thought-out. Proper etiquette with marathon-texts: Let the boy lead and try to end it first.

Text a boy to the "friend-zone"

Just not that into him? Be clear but polite when placing boys in the friend zone.

Just throw in a Pal, Sport, Kid, Buddy, Champ, or Dude into any of your texts to him and he'll surely know where he stands. Example: "Hey Pal, dinner sounds fun but I'll need to rain check!"

Overloading your texts with Redundant Questions.

Texting should be kept light and fun. Sending questions like: "What do you think of Chicago?" or "What happened last night that you couldn't talk about?" are no-nos. Might as well pin the kid down and shine a flashlight into his eyes. Don't ask things that will take time and space to answer.

Feelings + text = bad idea.

It's hard to convey tone over text. It's safe to say that the tone in the majority of text exchanges is light and sarcastic. Therefore, writing things like, "I really don't see where this relationship is going" or "Who was that girl you were with last night?" don't hold as much validity when said over text. Proper flirtextiquette is to not load your flirtexts with heavy statements/questions about your feelings. It's best and more respectful to discuss these matters in person or over the phone. If you already have trouble with anxiety, we strongly discourage

Don't be that girl.

We have noticed that some of our more popular girlfriends have become addicted and attached to their BlackBerrys, in an unattractive way. Whether they are texting, BBMing, or e-mailing, their cells are always in their hand and they're always typing away on them. Don't be that girl. There is nothing more annoying than trying to hang out with someone who is constantly directing her attention toward other friends wirelessly. That's being social by being antisocial. And don't even get us started on the irritating sound consistent tapping away on the keyboard makes while we stare at your forehead. It's a surefire way to turn any boy off.

you from discussing feelings and important issues over text. Waiting for a response text is an intense, emotionally driven situation. Sometimes we might text outlandish things just to get a quicker reaction out of him.

If you *don't* want a date, Abbreviate.

Listen up: the overuse of abbreviations, acronyms, and smilies in your flirtexts will put you on the next bus back to Single-town. If you want to avoid this be weary of how you write out your words and avoid sounding like you just stepped off the set of *Clueless*. While shortening words and letters is cute and funny when exchanged among your girlfriends, please don't use "brill" and "perf" in your flirtexts. They make you sound flighty. Constant abbreviations are totally girly and will make any guy roll his eyes—not laugh.

Obviously you are going to have to abbreviate some words in order to make your whole flirtext fit into one message. But sending "gr8 2 see u 2nite!" is not okay. Rule of thumb is that if you don't *have* to abbreviate your message in order to make it fit, *don't*. But if you need to cut down a few words so that it fits into one text, start by shortening words like "you" and "are" to "u" and "r." Catch our drift? If it still doesn't fit into one message, than turn that message into two texts, which is an appropriate solution.

The same goes for *overuse* of your smiley faces, "LOL"'s, and "haha"'s. The overuse of these makes you seem like an overzealous and hyper pre-teen girl on her way to a Jonas Brothers' concert. Let's reflect on these a moment:

:) — Okay, we know you think these guys are cute, but in reality they are annoying and a turn off when overused. So think twice before using them. Be aware that guys who overuse these may be slightly on the cheesy side.

Hahas and LOLs—We really like these and think they are cute ways of signaling to your crush's ego that he in fact is making you laugh. It's also a good sign when he sends you one, because he thinks you're funny. These laughing signals however tend to be abused in flirtexts. Just know that every semi-funny line he sends doesn't warrant a laugh from you. Stay true, be genuine, and pick and choose your LOLs. After all, he is going to be sending you all sorts of funny texts to try and impress, don't make it too easy and obvious for him.

the mis-text

texting tell-all

Always double-check.

"I just got off the phone with my ex-boyfriend, Jeff, to tell him that I was moving to his city. I totally played it cool on the phone and alluded to the fact that it was no big deal and maybe I'd see him around sometime, when really I was jumping off the walls excited to be moving to the same city as him. Literally two seconds after we hung up the phone, I text messaged my best friend, "Jeff just called! OMG he totally still loves me!" The second the word "sending …" appeared on my phone, I realized that I had sent that ridiculously cocky text to Jeff and not to my BFF. Lord help me."—Our very own Debra Goldstein

A mis-text is every texter's worst nightmare! Sending juicy information to the one person whose eyes it wasn't meant for spells T-R-O-U-B-L-E. Before you go nuts and throw your cell out the window (like Deb wanted to after her monumental mis-text), take a deep breath, and know that even though this may seem like the most embarrassing thing ever to happen in texting history, damage control is possible and just a text away.

In Debra's situation, she quickly bounced back and had no choice but to acknowledge her mis-text in the best way she knew how. Since both she and her ex–BF are movie buffs, she quickly retaliated with: "Uh, that was a movie quote!?" When he wrote back, "Passion of the Christ?" Deb knew she was back in safe territory. This fire was contained; however, she learned a tough lesson. At times when we're overcome with excitement, it's more likely that we can become careless with our texts. As caught up in the moment as you may be, it's always smart to double-check your text before sending.

Our grade school teachers said it best: "Remember to always double check your work before handing it in." Meant nothing to us then, but everything to us now. It seems like all technology these days gives you the ability to double-check your work before turning it in. Voicemail lets you listen to the message you left, with the option of erasing it and starting over. Digital cameras allow you to look at photos immediately and decide if you want to re-shoot. Text messaging only sends messages when you are ready to send. As a precursor, phones even hold spell check capabilities. You hold the power of pressing send. So make it your responsibility to double check your text, ensuring that all systems are go. Damage control is okay but prevention is best. Here are some proactive tips to avoid mis-texts.

TIPS TO AVOID
MIS-TEXTING

- Proofread all texts that would be considered risky if it fell into the wrong hands. ALWAYS check the name of the person who the message is going to before pressing send. It takes two seconds and can save much humiliation. Especially when you are gossiping. All it takes is a quick glance to make certain it is addressed to the right person.

- Rough draft your important messages in another mailbox first. If you are worried about what to write and need to draft your text a few times before sending (which is totally normal), do so in a different mailbox. Disaster strikes if you accidentally press send and you send the draft message, "wanna go to the *fart* party?" when you meant to say, "wanna go the *frat* party?" Simple solution: have your own name and number in your cell contacts and send the really special BPTs to yourself first, ensuring they read correctly. Then copy and paste the text to the lucky lad and rest assured he got the message right.

- Most cell phones have spell check capabilities. Use this option when sending important texts. It's a small thing that can make a huge difference.

Texts we deem important enough to draft in a separate mailbox:

1. Any text having to do with or being sent to a crush.
2. Texts revealing sexual encounters.
3. Gossip texts talking about other people.
4. Anything work-related.
5. All racy photo texts.

With cell phones getting smaller and technology more instantaneous, mis-texting has become all too common. We sometimes get anxiety about mis-sending a message to everyone in our address book. Being aware of the possible mis-texting situations will help you become proactive and avoid social suicide. The last person you want to brag to about the details of your date last night is your date. Yikes.

Here are some mis-text examples plus damage control texts. Our best advice to you when a mis-text occurs is to own up to it, find the humor in what happened, and pray for forgiveness.

Examples of Mis-texts and Damage Control

Mis-text 1:

Text meant for your mom: (following first date with new guy) "Best date ever! Chris is a keeper. Get started on wedding invites!"

Instead sent to: Chris

This is the most common kind of mis-text. It happens because sometimes our fingers and mind don't work fast enough and messages get mixed up. So even though you meant to text your mom, you texted Chris, because clearly he was on your mind. This can happen if you are not careful but it is easily avoidable.

Damage control:

"Sorry meant for my mom. Who you will be meeting at our wedding. Next week work?"

or

"Did you know that 1 in 7 texts you send are mis-texts? Looks like I got 6 more till my next:) P.S. I had a great time last night."

Mis-text 2:

Text meant for your BFF: "Guess what?! I went home with Dan last night! Call you later with the deets."

Instead sent to: YOUR ENTIRE ADDRESS BOOK. Including: your mother, your father, your boss, your grandmother (who hopefully doesn't know about text yet), and Dan!

Damage control:

"I'm happy to have provided you all with your Page Six morning gossip. Please don't get used to it as it is only a trial subscription."

or

"Oops. Well at least you guys all know how my Saturday morning's going;)"

What do you do when you're on the RECEIVING END of a mis-text?

Mis-text 1:

Current guy you're dating texts: "You looked hot last night."

Sent to: You. Which is odd since you DIDN'T see him last night!!

If your PBF texts you "You looked hot last night" and you spent last night at your parents' house on Long Island, then *ding ding ding,* we have a problem. Go ahead and entertain the elaborate story he's sure to make up about how the second part of his text got cut off, which was "... in my dream. In my dream you looked hot last night." Unlikely. Trust your gut on this one. Looks like you just caught him red handed.

Damage control:

"I think you meant to send this to your other girlfriend" *or* "Funny, I don't recall seeing you last night. But yes I did look hot."

Try to remain calm, cool, and collected. Always come out with your head high in situations like this. He should be calling you right away to resolve his detrimental mis-text.

Mis-text 2:

Current guy you're dating texts: "I took Molly home and hooked up with her last night."
Sent to: You. Molly.
Damage control:
"She must have been wasted." *or* "Wow you're one lucky fella."

Even though you may be upset, find the humor in it. He is bragging. He obviously thinks you are a catch. Yes it's somewhat immature that he is texting his friend about your make-out sesh but it's also a little flattering, no? Remember when John Travolta and Olivia Newton John's characters in *Grease* had way different reactions to their "Summer Lovin'"? She wailed about "staying out till 10 o'clock" while he made quips to his guys about "making out under the dock." Two very different sides of the story. Guys are cuddly and romantic with you and mean it but have a need to show off in a macho way to their friends. Best to teasingly call him out and then let it go.

texting under the influence (TUI)

Drunken texts are the pink elephants of our night owl existence. As much as you would like to

texting tell-all

"We all do it and live to tell the story the next day. It's the strangest thing … no matter who I'm out with, someone, after a few cocktails, always thinks it's the most brilliant idea to text something to their current crush or old flame. It's like an episode of *Sex and the City*. I've seen all forms of prevention for the drunk text: handing over your phone to a friend, signing up for a service that prevents you from dialing your crush's number past 10 P.M., making your friend swear not to let you dial anyone but your mom. The reality of all this is grim and short-lived. As you down cocktail number three, the chances of your not texting your crush are about as slim as Samantha vowing to become celibate. We all break after some point and think Carrie Bradshaw couldn't have written a better text herself." —Olivia

deny and forget the fact that you drunk-texted, you can't, because honey, that crap is cemented into your phone! The only thing worse than waking up to discover your text log full of outgoing flirtexts that you don't remember sending is reading those texts. They were most likely sent to a current crush or an old flame suggesting things you never would have suggested while sober. Not to mention they are probably filled with multiple misspellings and incomplete sentences, making you sound ridiculous. You are so embarrassed by these texts that you turn your lights off and go back to bed, hoping that when you wake up for the second time this would have all been a bad dream. At first you may not believe it, but trust us, we'll explain how there's life after a night of Texting Under the Influence ... or TUI.

The TUI phenomenon happens because alcohol lowers our inhibitions and impairs our rational thinking, making us more likely to say exactly how we feel without censoring ourselves like we normally would. And being that text messaging is instant and easy, it's our go-to when that boost of liquid courage gets ahold of us, making TUI all too common.

If you drink alcohol while in the presence of your cell phone, then sooner or later you are going to want to drunk text. It's human nature. We too have been guilty of TUI a time or two and know the effect alcohol has on our inhibitions. That is why we strongly recommend never, *ever* doing it. Below are explanations behind common drunk texting scenarios and some sweet tips on how to avoid TUI all together. And let's not forget the boys, also guilty as charged. In addition to helping you save face, we've also included suggestions on how to handle their late night texts and inappropriate booty calls.

Scenario 1: "I look damn good tonight."

You just had your eyebrows waxed, your hair looks the best it has all month, and your new DVF number is earning compliments all around. You're out with your friends, and think what a great idea it would be to text a crush (whom you haven't heard from in a while) to see what he's doing tonight. He's still on your mind and you think that tonight's the perfect opportunity to see him because, well, you look hot!! The alcohol has given you just enough courage to get out your phone and begin to compose a drunken text to the unsuspecting lad.

Truth: Step away from the phone there, Miss von Furstenberg! Yes, you look hot, but hey, guess what? Your crush can't see that. Word to the wise: Anytime past 10 P.M. is never the right time to be catching up with an old crush. It's 10 P.M. and the boy has already made plans. Plus, with your late night text, you've pretty much agreed to make out with him without him having to lift a finger.

Furthermore, your texting him after hours is a sign that you are swinging solo, Sister. Basically, due to this needy reach out, you've set the tone for the type of relationship you are willing to have. The last thing you want is to come off looking desperate. Avoid this at all costs.

How to Avoid Scenario 1: Remind yourself that you're out for a reason. You're drinking to

> *Think about it. If you had a hottie on your arm, the last thing you would be doing is texting an old flame.*

let loose, have fun with your friends, and forget about the stresses of work, bills, and the crises of your personal life (including long-lost BFs). Distract yourself from your phone. Go find a boy at the bar and make funny conversation to keep yourself entertained. This is a great distraction to get whoever else is on your mind off of it. A good laugh in the morning is always better than a good cry of humiliation.

OR

If you're really hung up on getting in touch with him, make a pact with yourself. Tell yourself that you will call him tomorrow during daylight hours. Planning something like a phone call to a crush can be exciting, plus when it's done appropriately—aka not in a desperate, drunken stupor—he will be genuinely pleased and impressed with your call. Additionally, this will a) give you time to see if you still want to text/call him tomorrow when you're sober (the true test) and b) prevent him from losing any interest in you because you texted him while inebriated.

Scenario 2: Can't let go of the past and move on.

You are out drinking and just ran into your ex-boyfriend's brother at the hotspot Eleven July where you and your FBF (favorite boyfriend) first met. All of a sudden, the DJ spins J.T.'s song "My Love" on his turntable. You begin to get sentimental because that was your song. As fate would have it, at that very moment you see a guy who looks just like your ex walk by. Your heart skips a beat as you pick up your phone to drop him a line about the crazy coincidences that just occurred.

Truth: In this scenario, you either just broke up with someone or never received closure on a very significant relationship in your life. You can't fathom him being with anyone else. You dwell on your past times together, and the mere thought of seeing his name pop up in your inbox makes your knees go weak. Bottom line, you still have serious feelings for this guy. There are so many things left unsaid, and the alcohol, mixed with all the coincidences, has made you want to text him. You miss him.

How to Avoid Scenario 2: Friends don't let friends TUI. Appoint one of your best friends as the person you always, we repeat, *always* text whenever you feel the desire to contact this FBF. Let this person know that whenever you get the desire to

texting tell-all

ALERT ALERT: DANGER ZONE. Not only are you drunk, which makes you ten times more emotional, but 1:30 in the a.m. is not the time to be opening up about your feelings. In the end you will only be putting yourself and him through more pain. We know it's hard to remember this at a time when all the signs point in his direction, but you're not in a relationship with him for good reason.

No matter how many good times you had together, it ended because one of your hearts wasn't in it. And at the end of the day, that's truly all that matters. There's absolutely no point in dragging this out any longer if he's not your Mr. Right. We know this isn't the answer you want to hear because your FBF is the one you hold a candle for. But sweetie, life goes on and you will rise again. We guarantee it.

text him, you will text her instead. Texting her will serve as your safety net for doing something you would later regret. In return, she will most likely remind you of his bad qualities and make texting him less appealing.

The logic behind this is that your friend acts as a sponge, allowing you to release any of those "loving feelings" on her instead of him. By texting her first you allow time to pass by when your desire was the strongest to text him. Hopefully by the time she writes you back your desire to text him is "gone… gone… gone… wooooooh." Plus, you most likely checked his Facebook status earlier that evening and know he's in Vegas for a Texas hold 'em tournament with his buddies. There's no way it was him you saw at Eleven July!

Scenario 3: Late night invites for drinks, a date, or make-out sesh.

It's 1 A.M. and you get a late-night drunken text from a PBF that says, "Whath arer you dooing?" You could be out on the town celebrating a friend's birthday or sitting at home watching *Oprah*. Either way, you have mixed feelings about his text. You're flattered that he is thinking about you, but at the same time, a little pissed off because this is clearly a drunk text. You like him, but are not sure how to respond.

In no way, shape, or form is this sort of invitation flattering. This actually *is* a cop out and if he had true esteem for you, he would have contacted you earlier. Trust us, we've asked all of our guy friends and the answer is always the same: Texts sent late night are purely sexual.

Like we said before, in life as well as dating, it's your responsibility to teach someone how to treat you. If you entertain his texts and late night pleas, don't be surprised if you start seeing a pattern of

> *Truth: Even though we are engaged in a texting culture, boys' manners shouldn't disappear because it's easier to approach you, nor should your morals and class give way to their easy plays.*

LNBT from this guy. That is why you have to take control from the get-go.

How to Avoid Scenario 3: If you like this guy and see potential for a future relationship, then ignore his LNBT. By not responding, you let him know that you're not one to give in to his drunken pleas and should be taken seriously. If he likes you, he'll take the hint and contact you the next day, most likely to apologize for his behavior. Be sure to let him know that you got his message but aren't the kinda gal to give in so easily. Next time he'll get it right and make pleas with you earlier in the day (and when he's sober).

Scenario 4: You texted him late night and he's not texting you back.

You spoke to him earlier and know he is out on the town tonight. He said he would text you later to see what you are up to. You're having a good time partying with your friends and into your second bottle of wine begin thinking about your crush, and checking your phone every five minutes to see if he texted. When you absolutely can't stand it anymore, you text him even though you know you shouldn't, to see what he's doing and he doesn't text you back.

Truth: This is a case of actions speak louder than words. He said he would text you but didn't. No matter what he may have said, his actions did not back up his words. And ultimately, his actions or lack thereof, are what you should be concentrating on. Such poor follow-up is a sign of disinterest. So by contacting him anyway, you are setting yourself up for rejection if he turns you down. Take his silence as a sign showing you where you stand. Know when to graciously bow-out.

The UGLY Truth: If you text a boy late night and he doesn't respond, he's most likely with another girl or doesn't want to see you. You can contemplate all you want about why he's not responding, but the bottom line is: He didn't want to make out. After all, he is a guy, and texting him late night is basically like offering yourself up to him on a silver platter. If he didn't respond, and you know he wasn't sleeping, take a tip and move on! He clearly has.

How to Avoid Scenario 4: Try to leave your conversations open-ended, allowing either one of you to get in touch. Meaning instead of saying, "Text me later" say "Let's keep in touch." This helps to avoid this kind of miscommunication altogether. Therefore, if you do text him and he doesn't respond, you can walk away guilt free knowing you did your part.

beyond
the basics

"Feels So Right, Yet Seems So Wrong": When Flirtexting is Cheating

"I can text multiple girls the same thing at the same time. Example, I can text 5 girls on a Wednesday the following: 'I was just thinking about how ridiculously good looking you are and decided I have to hang out with you this weekend—I'll do anything.' If only two of them are free, then I'm good to go Friday and Sat!"—Lane

Whether you're in an exclusive relationship or just dating around, you must be aware that flirtexting with other boys can create trouble in paradise. If you're in a committed relationship, any sort of playful text exchanges with boys other than your boyfriend is grounds for infidelity and will be held against you in the court of break-up. Whether you were only kidding around or serious, flirtexting others while in a relationship can be hurtful, especially if exposed unwillingly.

You single girls don't get off so easily. Even though you are single and flirtexting with many suitors, there are still guidelines to be followed if you're in the presence of other PBFs. There's truth

> *Out on a fourth date with Alex when you receive a text from Justin saying, "Hey hot stuff. Where u been? I believe I still owe you a drink . . ." Even though you like Alex and are totally into him, the flirt in you can't resist and you text Justin back an hour later with, "I believe you do. I'll let you know when I become thirsty!" and then continue your date with Alex.*

to every text. Below you will find the guidelines for what constitutes texting as cheating according to your dating status.

If you're single and dating . . .

You're a flirt who loves having adoration coming from several boys, especially when it's in the presence of one boy without the other knowing. It's a contradicting feeling. It feels so right yet seems so wrong. After receiving Justin's text you instantly get a cocky little grin on your face that puts this air of confidence around you, which you wear quite well the rest of the night. Something about being with one boy while texting another gives you a small rush that you love. And being that you're not tied down in a relationship you're pretty much a free agent who can trade up at any time. You've got a few boys vying for your attention and you're running with it. And, may we say, yay you.

The private and instant nature that texting allows makes it so easy to flirt with tons of boys at the same time, it's a crime not to. As serial flirts, admittedly this is one of the features we like most about flirtexting. It's an easy way to keep dibs on all your PBFs within a finger's distance. However, let us be quick to remind you, it's also very possible that any of these boys you're flirtexting could be flirtexting other girls as well. Enjoy it and take it for what it's worth: harmless fun that creates a little excitement.

Boys deep down know that single girls have several guys they're talking to at once (because they do the same with several girls). They may know it, but they certainly don't want to hear it or actually see it. So be mindful of a few flirtextiquette rules if the occasion occurs while on a date:

Rules of the Two-time Text While Out on a Date:

Don't leave your phone in plain view while on a date. Even though you're on a date with Jon but expecting a text from Jake, ensure your phone is placed in your bag and not on the table in plain view. Blinking red lights, vibrating sounds, and seeing messages pop up is distracting and will signal to your date that you've got better things to do than be there with him. Also, if you're not expecting a text from anyone, Murphy's Law will ensure when you're out on a date that other guy you liked will finally text you. Boys have a sixth sense and when you're out with another guy somehow feel the urge to contact you that very moment. It happens all too often, so play it cool and stow the cell.

Don't check your phone in front of your date more than once. Unless you're ready to explain why you need to check your phone, don't do it. Boys will call you out on this and have you explain why you're not devoting all your attention to them. If you don't want to be confronted, check your phone when he's not around. It is understood you're going to check your phone once during the date. Checking your phone incessantly displays disinterest. If you're really into a guy you'll refrain from checking your phone until after the date.

Don't text in plain sight and assume he doesn't know who you're texting. If he sees you texting he will assume you're texting another guy. It's just how his mind works. But then again, we would probably guess the same if roles were reversed. It's hurtful and can be uncomfortable to do this in plain sight. In general, if you're going to flirtext other boys while around your date please be sure to do so discreetly. While he's away in the restroom or grabbing your coat, these are open invitations to text another boy on the sly.

If you're having a sleepover with a boy, ensure your phone is turned off. If he's sleeping over and in the middle of the night your phone buzzes off the hook with texts, expect the boy in your bed to shy away shortly thereafter. Boys don't want to hear your phone buzzing at 2 A.M. They know it's not your mom. Checking your phone at night signals you're not interested and have needs to be entertained by someone else. It's courteous to check your phone in the morning. Most likely, you both will be doing the same thing.

If you're in a committed relationship …

Flirtexting other boys while you're in a committed relationship is not a light matter. Be it with an old boyfriend who you still keep in contact with or someone new, flirty banter over text is not cool. Text is the written word, so if anything, it is more real than if you say it to their face because it's traceable! Hello, text messages have forwarding capabilities so BEWARE!

If your current beau finds out you've been flirtexting with another guy, his feelings are going to get hurt. Use your best judgment and think through the consequences when responding to these texts. If you don't feel comfortable with your

texting tell-all

If you are in a committed relationship

Our friend Jessica and her boyfriend had been living together and were on the verge of getting married. They became friendly with another couple in their building, Ann and Chuck. Jessica especially became quite friendly with Chuck. A secret crush quickly grew between them and they began acting on this through covert texts. Before they knew it, these flirtexts described fantasies of being together. One day Jessica accidentally left her cell phone at home. Her boyfriend heard it buzzing and went to check her message. The message was a flirtext from Chuck saying "I want you now! Tell me again when can I have you?" Spotted: a heartbroken boyfriend who just caught his girlfriend red-handed. Update: Jessica is currently single.

BF reading them, then you shouldn't be texting in the first place. Better to be safe than sorry.

Below are the boys who probably still text you even though you are in a relationship and tips on how to handle it:

Flirtexting an Old Boyfriend

Look, we know that you once had feelings for your old BF, and it feels good when he texts you out of the blue. However, getting in touch every once in a while is very different than text messaging every other day. After all, this is an ex-boyfriend whom you once had feelings for. Even if these flirtexts are platonic it will still be very hurtful if your current boyfriend finds out. Using the silent treatment in this situation is universally known as a way to get out of verbally stating the obvious: that you're "with someone." He should take the hint and stop texting you. If he doesn't, then text him: "gonna have to call it quits with the flirtexting for now. The new Boyfriend might get jealous. I'll alert you and the press if things change."

Flirtexting with Guy Friends Whom You May Have Hooked Up With

These are the guys that fill up your little black book and who you probably flirtexted with before you were in a committed relationship. They come in and out of your life and check in every now and then. They could be aware that you have a boyfriend but that doesn't stop them from going after some attention from you via text. If they know you have a boyfriend and are still pursuing you, it's not cool and their attempts should be shot down. If they don't realize the degree of how "off-limits" you really are, make it clear and text: "Sorry bud, you're a few months too late. I'm a taken girl now."

Flirtexting with a Guy Best Friend

This is tricky territory, as he is considered one of your friends, right? Boyfriends in general don't feel comfortable having you be so chummy and flirty with another straight guy. Even the most secure guy will still be on edge if this is the case. They know how their own minds work and assume this guy friend has more than "friends" on his mind, which they may not be completely off on. They may tell you they are cool with your close friendship with this person, but they still look at him as a potential threat to your relationship. We advise you take your boyfriend's feelings into consideration when sending flirtexts to your guy best friend. Even though he is your "friend" and sending suggestive flirtatious texts to one another is what you've always done, its still grounds for jealousy and accusations of cheating. It's best to keep your responses light and clean to avoid any problems. To set the record

> **FLIRTING:** The harmless act of being affectionate toward someone in order to make them feel good about themselves, without necessarily having an emotional commitment to them. You could just be a friendly girl who chats throughout the day with the coffee barista, your coworker, your professor in your favorite class, you get the idea. This kind of flirting is acceptable and welcome. However, there is a difference between harmless flirtexting, and pure, outright, hitting on someone over text. Be aware of the difference and act accordingly depending on your relationship status.

straight, try texting: "You know I love you but we've got to take a time-out with these flirtexts. I'm not in the market for a jealous boyfriend. We can open up the floor again if this one doesn't work out;)" Your best friend should understand because he wants what's best for you and therefore will respect your relationship with your boyfriend. And your boyfriend will continue to keep his mouth shut about your close relationship with another guy.

Textual Intercourse

"My flirtexting has dramatically decreased since getting engaged. My flirtexts now consist of messages like, 'Babe, do you want me to pick up a chicken for dinner?' Although I'm fairly certain that turns her on."—Scott

how to spice up an existing relationship through texting

If you are reading this chapter it means that you have successfully passed through the early stages of flirtexting and into a fabulous committed relationship. A congrats is in order! Now, just because you have entered into the land of "we" does not mean that you can call up your cell phone provider and cancel your unlimited texting plan. Flirtexting is still fantastically important in an existing relationship—only now you're able to take it one step further.

In a relationship, the fact that you both care deeply about each other is out in the open and felt equally by both parties. This commitment allows you the luxury to relax a bit more when sending texts. You don't have to spend as much time on your BPTs and you can go ahead and throw texting timelines out the window! Whew. Flirtexting in a relationship opens the door for you to explore the other, shall we say, more *risqué* things you can do with text. With that, feel free to text things that only prove how into him you are by *going beyond the flirtext*.

sexting

"I like texting because you can turn a girl
on via text while you are on your way to
her house, telling her just how you want her
upon your arrival."—**Evan**

When you've been with the same person for a while, what better way to spice up your relationship than with a fiery flirtext? One that will arouse more than just his senses (wink wink). Yes that's right. We're talking about the sext-text, or sexting, and it's a long-term dating DO! It's sex over text, and it's HOT!

> **Sexting:** Descriptive sexual conversations through text with the intention of arousing the person on the receiving end.

When the mood strikes, you can reveal fantasies and sexy thoughts to him while he is out of town, at work, or even across the dinner table. There's always time for a fantasy text!

Examples:

- While the two of you are at dinner with friends, text him that if he casually uses the word "wet" (or any funny/sexual word) in conversation at least six times before dessert comes, then you'll do that thing he loves later.
- When you're at dinner and he gets up to use the restroom, text him "did I mention I'm not wearing any underwear?" (even if you are). He'll be asking for the check before you order your main course.
- If he's on a business trip and you miss him, text "I'm all alone in this big bed of mine. What should I do?" Believe me, he'll gladly take it from there!
- If your man plays sports and you are at this game, text "Score now and I'll let you score with me later;)" or "Win or lose, you're scoring with me tonight;)." If he checks his phone

during halftime, watch as he runs faster, hits harder, and throws further during the second half. If he doesn't get the message till after, he'll still appreciate the flattering gesture, win or lose.

• If he asks you "Want to go to a baseball game?" respond with "Sure, but only if we can go to third base;)" OR if he says, "Basketball game tonight?" you say "Are you insinuating foul play? Love to."

Long-distance Relationships

Sexting is especially beneficial in long-distance relationships. The need for sexting heightens when you're not around one another. It's an excellent way to remain close, especially if there's a time difference. If you are getting ready for bed on the East Coast and he is finishing up a meeting on the West Coast, he is still able to connect with you in your time of need. Sure, it's not the real deal, but hey, it beats nothing at all! (Um, did we mention this is a great form of safe sex?) Texting helps make long distance more bearable by having the ability to be in constant contact.

Virgin to sexting? Not to worry. It doesn't hurt. Talk about that thing he did to you last night that you loved. Text what you want to do to him when you see him next or what you are dying for him to do to you. Talk about how your salad came with a really big cucumber and you thought of him. Be bold, be blunt, and be bad. Be a little selfish when you dirty text and say things that will turn you on in the meantime. Chances are if it turns you on, it'll turn him on twice over.

Shy by nature? This is a great way for those of you on the shyer side to let your man know what you like sexually. If you are too embarrassed to tell him in person, use text and the casual and safe environment that it provides to tell him your fantasies. You'll be surprised how much you will benefit from being open about your likes and dislikes through sexting.

Not into sexting? Flirtexting with your man doesn't always have to be dirty. Texts like suggesting pizza and football for a Monday night date or randomly sending "I love you" go a long way too. Long-term relationships can get mundane so you've constantly got to be doing little things to keep the sparks alive and the butterflies flying! Sending a random, thoughtful flirtext is a great way to show him how much you care.

- Randomly text, "I love you" or "I'm crazy about you." It's simple, it's quick, and it will make him smile.
- Send a recent sports fact about his favorite player like, "Ole Chipper just hit his 400th home run last night, eh?"

- If he's having a rough day at the office, suggest a date night including his favorite things and tell him about it to get him through the five o' clock whistle. "Agenda for the evening: you, me, your favorite Philly cheesesteak, a Heineken, and the Red Sox–Yankee Game. See you at 6."

With technology these days, the sexting possibilities are endless. Imagine what you can do if your cell phone had a camera on it. Oh, but wait . . . it does.

Photo Text

Ah, the infamous photo text. Now that most cell phones have cameras in them, sexting has been taken to a whole new level. All those scandalous things you're texting your man can now include illustrations and pictures. WOW! The only thing better than sexting is a sex text that includes a classy, yet sexy photo starring yours truly!

Men happen to be visual creatures. Therefore, we understand the enthusiasm that guys are overcome with when it comes to the camera phone. We're keen on taking a little off for his entertainment pleasure, but he's going to have to be involved to see the real deal.

Whenever you are taking a racy photo text of yourself, proceed with caution. Never include your face in the photo. Even though you are in love now, what happens if something goes terribly wrong and you have a nasty break up later? You guessed it! Hello Facebook photo tagged of you and your large breasts! May we remind you that you're also friends with your boss? Proceed with caution.

Here are a few classy ways of using the photo text to spice up an existing relationship:

- Take a picture of you sticking your head out of the shower, showing some sexy shoulder or your smooth leg, and write "can you get me a towel?" or "bath time would be soo much more fun with you...."
- Take a picture of your bed—or better yet, you in it—and write "room for two" or "Can you meet me here later?"
- If you bought some new lingerie and are planning on wearing it that night, take a picture of the lingerie still in the Victoria's Secret bag and write "I bought you a present ... if you're a good boy, I'll give it to you later." He'll be so excited he's sure to be on his best behavior.
- If he tells you to take a photo of yourself and you don't want to, take a picture of your shoulder and write, "guess what part of my sweet body this is from?"

- Send a snapshot of your lips, with a message that reads "they miss you."
- If he's begging for you to take a sexy picture of yourself to send him, take a picture of your friend's cleavage and write, "Not mine, but I'm still thinking of you;)"
- If you're on a beach vacay and missing your man, snap a pic of you in your hot bikini and write, "wish you were here" in the sand.

WARNING ABOUT USING THE PHOTO TEXT:

If you are NOT in a serious relationship with a guy and you text him dirty photos of yourself, he will forward them to his friends. We can't stress enough how you must

> *Guys can lose their heads, and their phones sometimes, and we don't want to see your boobs on the Internet.*

proceed with caution when sending racy photos of yourself. So be smart, ladies, and trust us on this one. Boys can be pigs sometimes. The last thing they are thinking about when they receive a hot photo of a *new* girl they are dating are her feelings. We've seen too many of our girlfriends get hurt because they sent nude pictures of themselves to guys they were "just dating" to find out they've become the screensaver on his friends' phones. This is why we only advocate sending these kinds of texts when you are in a committed relationship. Otherwise, expect what you send to him to be forwarded to his entire soccer team.

Flirtexting Game: Scavenger Hunt

Heart game: Before you go out that night with your BF, color ten hearts on your body with a red Sharpie. Make some obvious (inside of wrist, on finger, on back of shoulder) and others hidden (behind neck, on inside of thigh, behind ear). Right before you get to dinner tell him that you are playing a game with him tonight. Explain that you have ten hearts on your body and over the course of the night he has to text you places where he thinks they are located. If he guesses them all by the time you are ready to head home, then his prize is that you will strip down showing him exactly where each is located on your body. (You can also add that he can kiss each location while you are showing him.) Feel free to give him hints if he's struggling (example text: "I go crazy when you kiss me here").

This game only works when you are planning an evening out with friends or other people. There is something super sexy about having a steamy conversation in front of other people that only the two of you are in on.

The most important thing to remember when using text to spice up an existing relationship is to do a little bit of everything. Send him dirty ones when you're feeling a little feisty, fun and playful texts when you're thinking about him throughout the day, and intimate flirtexts when you miss him. The key is to mix all these different kinds of spices together. This will keep him on his toes and wanting you more and more with every flirtext sent.

texting tip

Facebook, BBM, E-mail, and Other Non-Texts

"Texting is the best b/c there is no pesky 'BLOCK' feature like on IM. The only way to stop a text is with a restraining order … and who has the time to go thru the hassle to get one of those?"—John

Text messaging is the most widely used and recognized tool in dating today. It begins and ends with the text message. Its universal accessibility and popularity has proven staying power greater than any technology fad. We do however find it necessary to acknowledge our other techno-dating and-relating friends. Modern times have supplied us with a few supplements that we found also enhance our social calendars. Enter stage left: e-mail, BlackBerry Messenger (or, as the cool kids say, BBM), Facebook, MySpace, and instant messenger or IM. These are all ways we stay in touch, network, and of course, flirt. From a dating standpoint, determining which avenue a boy approaches you says a lot about them. Read below to see what each means.

BBM: Stands for BlackBerry Messenger. It's portable instant messenger for people with BlackBerrys and it's free. If you can believe it, BBM is even more instant than text messaging. It works like instant messenger for BlackBerrys. It shows when someone is typing you a message and confirms when someone has read your message, adding a whole new component to communidating.

Pros:

- Faster than text messaging.
- It lets you know when messages are read, leaving no question that they didn't get it.
- Exclusive to BlackBerry users.
- You have to exchange pins in order to connect.
- Free.
- Fun!

Cons:

- Exclusive to BlackBerry.
- It tells you when someone has opened and read your message, making the excuse "I didn't get your message" not an option here.
- Adds an element of guilt. It's a diss if someone reads your message and doesn't respond.

texting tell-all

BBM is just like a text!

Secret unveiled: How to check a BBM without getting the "R?" Someone BBMs you and you want to read it but don't want them to see that you read it, aka the "R" you get once you know someone read your messgage. Here's what you do: you go to another BBM conversation and wait till the bottom bar notifies you that Dave wrote you "Hey what are you …" and you can't make out the rest. You roll over this part and right click "copy"— go into a blank message box and "paste." Your entire message Dave sent you will be pasted without him ever knowing you read it. Now rest assured you have all the time in the world to write back. You took the time clock off a BBM which makes it a text again. You're welcome.

texting tip

Now say you met a boy the other night and didn't get a chance to exchange phone numbers. You most likely will receive a Facebook alert inviting you to become his "friend" the following day. That is, if you don't beat him to it. From there, a friendly poke or message trail begins that can lead to exchanging numbers, an official date invite, or joining his Facebook entourage! Facebook has alleviated the stress of shoulda coulda woulda asked you out, because before you know it, you've been "Facebooked" (a term widely used among Facebook faithfuls). All you need to know in order to find a boy on Facebook is his name or a mutual friend; the rest involves some digging on your end.

Facebook: This started as a social network for college kids and has recently grown and expanded. From students to business professionals to celebrities, everyone who knows what's good for them has joined the ranks of the largest social network today. You create profiles according to your "network," which consists of the school you attend/ed, the city you reside in, and the job you have.

This makes for an instant connection between people who went to the same school, are from the same town, or work at the same corporation. The beauty is you can put as much or as little information as you deem necessary on your profile page. If a PBF checks out your page and likes what he sees, expect a friend request and message soon thereafter. Facebook is also a great way to keep dibs on any boy by checking his status (where he is, who he's dating, what he ate for breakfast), recent photos, wall posts, and mutual friends.

Pros:

- You can read up about a person and their whereabouts based on the public information that is readily available thanks to photos and status updates.
- Before accepting someone's friendship request, you can check to see what friends you have in common. This will either help you remember how you met, or send up a red flag that they are in fact a stranger. (Remember, stranger danger!)
- You can set your profile to private so that only people in your network can view your page.

- If your friend wants to set you up with someone, you can check their profile to see if you are interested before committing to a date.
- You can spy on ex-boyfriends or people from your past.
- You can chat with people worldwide for free.
- You can Instant Message your friends if they are logged onto Facebook at the same time you are.
- If you're browsing your friend's page and stumble upon a hottie who is PBF quality, all you have to do is ask them to be your friend and watch as your relationship blossoms! No harm in that!
- You can rekindle friendships with people whom you have lost touch with.
- Facebook will suggest people that you may know or want to be friends with based on your current friends and networks.
- You can add the Facebook application to your BlackBerry, so that you can quickly detag any unflattering photos before anyone notices.

Cons:

- Snooping around on Facebook can sometimes cause you to find out more than you bargained for. Like when that boy you like changes his relationship status from "single" to "in a relationship."

- Your friend could "tag" you in an unflattering photo for the world to see.
- People know when you don't accept them as a friend. It's a snub.
- It's addicting.
- Employers have taken to looking at someone's Facebook page before hiring employees. It's a way for them to check out the real you. Be sure not to put anything unbecoming up. You never know who is looking.

MySpace: The first and still widely popular online social network. Open to anyone and everyone who has Internet access. It started as a way to help people connect to those they have lost touch with. If you don't have their number anymore, chances are you can find them on MySpace. You can also see what they've been up to since second grade (i.e., where they live, what they do for a living, if they're married, or if they've gotten fat!). A lot of lame boys also use MySpace to pick up girls. For some reason they think that if you are on MySpace it is okay to hit on you, but hello, if we don't know you it's creepy.

Pros:

- It's the biggest and most popular social network, making it likely you will find who you are looking for.

- You get in touch with people who you have lost touch with.
- You can keep up with what your old boyfriend is doing these days.
- There is a safety feature allowing you to block strangers from looking at your page.

Cons:

- Because it is so widely used, there is a lot of unwanted solicitation from strangers, bands you've never heard of, and people from your second grade class that you don't care to rekindle relationships with.
- Everyone who once had MySpace, has now moved to Facebook and therefore hold two domain homes. This makes it hard to keep up with which one to use to contact them.
- People can call themselves whatever they want as their user name, making it difficult to find people.
- If you can't find them through their user name, you have to have their email address to find them.

E-mail: More often these days, people prefer to give out their e-mail address, instead of their phone number, to someone they just met. E-mailing is a more neutral means of connecting that doesn't blatantly read "I want to date you!" This form of communication is most popular amongst the thirty and up boy bracket. Boys who e-mail are usually those who didn't have cell phones in college and at one point relied on e-mail to feel out a girl's interest before asking her out. It is not uncommon to be asked on a date over e-mail and to accept.

Pros:

- E-mail allows you endless space to type your message.
- If you're not into him, the excuse "I didn't get your e-mail" is most believable here.

Cons:

- Your e-mail can get lost in SPAM and you will never know.
- Some of us don't have e-mail hooked up to our phones so it requires you to be constantly on the lookout for a computer.

Pin: Also exclusive to BlackBerrys, this is your individual code you use to begin a BBMing dialogue. Alternately, you can send someone a message, much like a text, via their pin. Meaning you can use their pin as their number and send a text that way. You are still able to see whether or not that person read your message and all of your past messages to that person will show up in the log. This is a less popular method, though used more frequently in the case that someone doesn't want to commit to the time sensitive pressure of BBM.

Pros:

- You can tell when the person you sent your message to read it.
- You can see all messages sent to that person on one screen.
- A way to connect without exchanging phone numbers.

Cons:

- It's not well known and therefore not as commonly used.
- It's a little aloof. Getting a pin from someone can be awkward because it is so infrequently used.

Instant Messenger or IM: IM is a way to have quick conversations with your friends over the Internet. Usually when you log onto your e-mail account, you are able to instantly IM with other people who are logged on. Don't be surprised if while you are checking your e-mail, you get an IM from your crush, who is most likely also checking his e-mail or stalling from working on company time. As it was most popular in the nineties, the use of IM is dwindling and giving way to other techno-dating methods. However, once upon a time, boys would ask you on dates, express their love for you, and madly flirt over IM. Following the three-hour telephone conversation era, boys and girls resorted to "IMing" from their home computers because most likely they were writing a paper and enjoyed the sideline conversation that allowed them to "take a break." Since dating has gone mobile, so must the ways we date. Instant Messaging is no longer the way to a girl's heart. It's just used for quick sideline conversations.

Pros:

- Fun distraction when you're at work or writing a paper at the library.
- Quick, easy, and free way to say hello.

Cons:

- Was a late nineties fad. People don't use as often when it comes to communi-dating.
- What you write isn't taken as seriously as texting.
- If one of the parties logs off their email then the conversation will end without warning. If you weren't the one logging off then this could leave you thinking that it was your bad joke that made him end the conversation without notice.

bonus features

Movie Quotes

"It's easy to drop fun little hellos and stupid little flirty stuff without taking the time and effort to talk on the phone, which I hate doing."—Adam

Crush: Party at Matt's house tonight. You in?

You: No. I've got my little sister's dance recital to go to.

Crush: Lame.

You: You better lock it up!

Crush: No, you lock it up!

We're not sure why, but boys love to quote. A movie, TV show, comedian, you name it. If someone else said it, they'll quote it. At times, we've even seen guys speak to one another solely in quotes, a phenomenon that can be rather annoying. But for some reason they seem to think it's hilarious, and to be honest, we think it's pretty funny too. There's nothing sexier than a guy with a great sense of humor, and if he's quoting funny movies, we betcha he falls under that category.

Boys like movie quotes and find nothing hotter than to receive one via text from a girl they are crushing. When used properly, movie quotes are a fantastic way to respond to his flirtext. Text messaging with boys is all about the *zingers*. Movie quotes combine the power of wit and useless knowledge and are sure to spark tons of intrigue from the lucky lad on the receiving end. Be it an old school movie or today's blockbuster hit, catchy, clever movie quotes all quickly assimilate into an average boy's daily banter. So when you fire back with a line from his favorite movie, watch as your dating stock rises.

Oh, and did we mention it's a surefire way to grab his attention, especially if he's been MIA recently? Boys have told us that when a girl nails a movie quote, it's super impressive and piques their interest even more. They get as much of a reaction, if not more, than that hook-line-sinker-fish move they go crazy for on the dance floor.

The appeal of movie quotes is exponential. Now, let us teach you a few tips on how to use them over text. Follow these rules and use some of our examples below and he'll be uttering "Here's looking at you, kid" in no time.

1. **Follow our adage: When in doubt, quote.** If he's not texting/calling you and you are thinking about him but have nothing of substance to say, instead of just writing "Hi," write a movie quote. What's more appealing to a man than a snappy one-liner taken out of his favorite blockbuster? Not much besides the fact it's coming from you, Missy!

2. **Ensure the timing of a movie quote makes sense and is therefore aligned with your deliverance.** Be sure that your timing is appropriate for sending the quote. If used in the wrong context, he might not get it and your attempt could backfire. Which means, "Houston, we've got a problem." Best time to flirtext a quote is when you recently saw the movie together or were recently discussing a movie you both like. Same goes for TV shows that are both your favorites and a new episode just aired.

3. **A word to the wise: be tactful with your quotes.** Just because boys love them that doesn't mean you should be throwing them around left and right. *Don't* overdo the quoting because it'll lead him to say "Hasta la vista, baby" and then we'll be saying, "Welcome to Dumpville ... population, you." Catch our drift?

Best case scenario:

You quote without using quotation marks, he gets it, and quotes back from that same movie. If you want to keep going with it and need more lines from that movie, just go online. IMDB.com has tons of quotes from every movie ever made. You can keep this dialogue going for hours of useless fun.

4. **When in doubt, use quotation marks.** We recommend texting your quote without using quotation marks around it. It's funnier that way. But if you think he might not get it, then go ahead and throw those suckers around your quote.

Here are some quotes that are pretty safe to use no matter what the circumstance. Take note that you don't have to use the whole quote every time. Just pull from it what you need that makes the most sense in that situation.

great movie quote texts

Argument

Use these when you are arguing with him and need some fighting words:

- "Erroneous! Erroneous on both counts!"
  —Wedding Crashers
- "Let me tell you a little story about a man named 'Sh.'"
  —Austin Powers
- "This is one doodle that can't be undid, homeskillet."
  —Juno
- "I'm not your buddy, friend."
  —South Park
- "Agree to disagree. When in Rome."
  —Anchorman
- "Hi, I'm Earth. Have we met?"
  —Tommy Boy
- "Put a cork in it, Zane"
  —Zoolander

Attention Grabbers

Use when you want to say hello, but don't want to just write "hi":

* *Your text:* "It smells like updog in here."
  Him: "What's up dog?"
  You: "Oh, nothing much, what's up with you?"

  —The Office

* "Bueller? Bueller? Bueller?"

  —Ferris Bueller's Day Off

* "I don't know how to put this but I'm kind of a big deal."

  —Anchorman

* "Hey Peter. What's happening?"

  —Office Space

* "I've been performing feats of strength all morning."

  —Seinfeld (Frank Costanza)

Great Random Quotes

Use when you need a great response back to his random text:

* "That's what she said."

  —The Office

* "I got to get outta here, pronto. I got a stage five clinger. Stage five, virgin, clinger."

  —Wedding Crashers

* "*Mortal Kombat* for the Sega Genesis is the best game ever."

  —Billy Madison

- "First rule of Fight Club: You do not talk about Fight Club."

  —Fight Club

- "The greatest trick the devil ever pulled was convincing the world he didn't exist."

  —The Usual Suspects

- "Do you ever get down on your knees and thank God you know me and have access to my dementia?"

  —Seinfeld (George Costanza)

- "Don't forget to bring a towel"

  —South Park

- "Looking at cleavage is like looking at the sun. You don't stare at it you get a sense of it and then you look away."

  —Seinfeld (Jerry)

- "Funny, she doesn't look Druish."

  —Spaceballs

- "I've negotiated my butt off, Giorgio.

  —Zoolander

- "Who is this? Uncle Leo?"

  —Seinfeld

Words of Encouragement

Use when he needs a little pick-me-up:

- "You're on the rebound. You're like an injured young fawn who's been nursed back to health and is finally going to be released back into the wilderness."

    —Old School

- "What are you going to do for an encore? Walk on water?"

    —Wedding Crashers

- "I'm sick of following my dreams. I'm just gonna ask where they are going, and hook up with them later."

    —Mitch Hedberg (a super funny comedian)

- "My advice to you … is to start drinking heavily."

    —Animal House

- "We got no food, no jobs, and our pets' heads are falling off!"

    —Dumb and Dumber

- "You stay classy, San Diego."

    —Anchorman

- Elaine: "Ugh I hate people." Jerry: "Yeah they're the worst."

    —Seinfeld

Weather Reference

Use when he asks you how the weather is:

- "It's so damn hot. Milk was a bad choice."

    —Anchorman

- "I'm sweatin' like a Tijuana whore!"
  —The Break-Up
- "It's too damn hot for a penguin just to be walking around!"
  —Billy Madison

Feeling Sick

Use when he asks how you are feeling:

- "I've got a fever and the only prescription is more cowbell."
  —SNL
- "I think I've got a case of the black lung, pop."
  —Zoolander

Travel/Airport

Use when one of you is traveling:

- "Why you going to the airport? Flying somewhere?"
  —Dumb and Dumber
- "I'll be back before you can say blueberry pie."
  —Pulp Fiction
- "May the Schwartz be with you."
  —Spaceballs
- "Are we there yet are we there yet are we there yet"
  —The Simpsons
- "There's no reason to become alarmed, and we hope you'll enjoy the rest of your flight.

By the way, is there anyone on board who knows how to fly a plane?"

—Airplane

Calling him out

Use when he is getting on your nerves and you want to give him a hard time:

- "I'd like to be cowboys from Arizona or pimps from Oakland but it's not Halloween. Grow up Peter Pan, Count Chocula."

  —Wedding Crashers
- "It's such a shame. You're so money and you don't even know it."

  —Swingers
- "Just when I thought you couldn't get any dumber, you go and do something like this and totally redeem yourself!"

  —Dumb and Dumber
- "You think that you are too cool for school, but I have a newsflash for you, Walter Cronkite... you aren't."

  —Zoolander
- "Fat drunk and stupid is no way to go through life, son."

  —Animal House

Admitting Fault

Use when you messed up and want to acknowledge it:

- "I'm sorry I called you a hillbilly. I don't even know what that means."

  —Wedding Crashers
- "SAMSONITE! I was way off."

  —Dumb and Dumber
- "Damnit Derek, I'm a coal miner, not a professional film or television actor"

  —Zoolander

Asking Him Out

Use when you want to see him and want him to know it:

- "Tower, this is Ghost Rider requesting a flyby."

  —Top Gun
- "I'll be in the neighborhood later on, and I was wondering if maybe you wanted to get some frozen yogurt, or perhaps a whole meal of food, if that would be agreeable."

  —Old School

The Language of Text

"Because when I'm with my friends, I can say cute things to a girl without having my friends give me a hard time."—Blake

Remember when "TGIF" was the only acronym you knew? NBC even dedicated a whole Friday night to it! Good times. Today, it's nothing out of the ordinary to shorten our words in order to save time and much needed space. Our busy lives have affected our language as well as the way we communicate. Proper English has transformed itself into abbreviations, acronyms, and slang. Through texting, we've officially created an unofficial new language: the Language of Text.

The language of text consists of short, abbreviated letters and numbers, combined with complete words. Text messaging or SMS allows for only 160 characters per text message. Since this is a limited amount of space to work with, wireless users created a language to accommodate text messages. Below is a dictionary for common texting lingo as well as some hip new terms we use in our everyday lives. Feel free to spread the word....

P.S. Our list is limited because we do not recommend going crazy with the abbreviations. Boys get turned off if girls use too many of them in a flirtext. But we urge you to adapt these and text away with your best girl friends. They are meant to be fun supplements to your texting conversations, not to be taken seriously. Just avoid looking like you spend too much time with your ten-year-old sister is all we're saying.

P.P.S. We wanted to note that a texting abbreviation longer than five characters long is not an abbreviation—it's just ridiculous. No one understands what you are talking about, and you run the risk of looking foolish so just don't do it. Below we've listed a few of these ridiculous abbreviations that prove our point:

AWGTHTGTTA: Are we going to have to go through this again?

ROTFLOLAPIMP: Rolling on the floor laughing out loud and peeing in my pants

MTFBWU: May the force be with you

flirtexting dictionary

<3 Love or heart

B4 Before

BC Because

Besos Means "kisses" in Spanish. An alternative to "xoxo" when signing off.

BF Boyfriend or current crush

BFF Best friend forever

BFFAE Best friend forever and ever

Boyf Boyfriend

BPT Best Possible Text. A formula created by sending just the right text, at just the right moment, to get you exactly what you want. These are the type of flirtexts you should always be sending to your PBF.

bro Brother

BRB Be right back

Brill Brilliant

BTW By the way

clutch When something is convenient or handy. Example: "My bf loves watching *Gossip Girl* too! How clutch!?"

ciao Can be a greeting or sign-off; means both hello and goodbye in Italian.

Cheers A British signoff.

C See

Dreamy The new "cool."

Epic Something that's a huge deal or great.

F2F Face to face

FBF Favorite boyfriend. This is the boyfriend that you liked the most, maybe even loved. We don't just throw this phrase around. There can only be one of these.

First Class Great

Fo Sho For sure

French Smooch, kiss, anything involving a little harmless tongue action.

FYI For your information

G1 Good one!

G2G or GTG Got to go

GF Girlfriend

Goss Gossip

IDK I don't know

JK Just kidding

KIT Keep in touch

KMA Kiss my ass

K Okay

LMK Let me know

LNBT Late Night Booty Text; a flirtext sent any-time past 10 P.M. asking you what you're doing. These are purely sexual.

LOL Laugh out loud

LOML Love of my life

Luv Love

Lylas Love ya like a sister

Mental Better than amazing. Example: "Um, did u see my new Louboutin's ? They're mental!!"

MTE My thoughts exactly

MYOB Mind your own business

NBD No big deal!

Neat The word we use when we're bored with what you're telling us. Example: Boy: "I make over 500K a year and drive a Mercedes and have a house in Miami … " You: "Neat." (You may or may not want to also roll your eyes here.)

NOYB None of your business

NRN No response necessary

Obvy Obviously

OMG Oh my god

PBF Potential boyfriend

PDA Public display of affection

Perf Shortened version of "perfect." Also to be used as a confirmation to something: "Meet me at 8pm?" "Perf."

QT Quality time

R Are

Rad Another word for cool or awesome.

Shut up! Get out!

Special To be used when someone's off the wall. Example: "He got kicked out of gym for wearing girl shorts. Yeah, he's special."

SOL Shit out of luck

SOS The international distress signal meaning "Save our Ship." We use it to signify "Help me."

SS Social Suicide. Example: Mis-texting your current crush (instead of your BFF) that you think his brother is hotter than he is would be total SS!

THX or TY Thank you

TMI Too much information

TTYL Talk to you later

U You

U2 You too

UR You're

Vintage Referring to someone you consider way too old to date, or a reference to old news.

Example: "Ryan is hot, he's just way too vintage for me."

Word Multiple meanings: for sure, I agree, absolutely. Example: Boy: "Meet me at my place at nine?" You: "Word"

WTD What's the deal?

WTF What the f★ck!

XO kisses and hugs

Y Why

Acknowledgments

We want to express our deepest gratitude to the following people, whose talents, encouragement, love, creativity, and passion have made up what we like to call: TEAM FLIRTEXTING.

To our family, for never doubting our dreams. Thank you for your tireless efforts of raising us to have high standards, so we don't fall for every guy who sends a good flirtext.

Chris Berkenkamp, our Consigliere. Your expert advice kept us at home writing instead of out gallivanting the city. Thank you for your guidance and for keeping us on track!

Biggest thanks goes to our incredible agent, Sarah Dickman, whose support and cheerleading is what has made our book possible. You believed in our project from the beginning and because of you, the world will now know how to flirtext.

Ann Treistman: We are so thankful to have such a cool editor who helped turn our flirtexting ideas, stories, and woes into a beautiful and easy-to-read guide. Thank you for your patience, help, and all your hard work in making our book the best it can be.

Thanks to everyone at Skyhorse Publishing for believing in us and for working so hard to get our book out there. We are truly grateful for all you have done!

LeAnna Weller Smith: Thank you for making our book so stunning. Your designs brought us to tears.

Maria Ponce: You are a rock star photographer with ah-mazing talent! We are so grateful for all your beautiful photos. Thank you!

Our friend and PR pistol Melissa "Little Memphis" Baer. Thanks for looking out for us and for promoting FLIRTEXTING where it counts. Love you.

Matt Sugarman: Thanks for doing a sweet job of making sure all our boring legal documents are in order. You are awesome!

Katie Dox: Thanks for your help beautifying our first born idea and packaging it well.

A special thanks goes out to a few of our incredible family and friends, whose involvement in our book has been invaluable. You guys helped us "shut it down": Courtney, Jenna, Erica, Candice, Mirelle, Peter, Lauren, Becca, Nisha, Anna, Jessica, Hillary, Jason, Marc, Nat, and Richard. We love you and are eternally grateful for your enthusiastic encouragement and generous support.

And to the rest of our wonderfully social friends who poured out their stories, life lessons, and amazing quotes to us, thank you! You all made this book better than we ever thought it could be. We heart you and owe you a beverage of your choice.

And finally, a great big thanks to all the hot boys of NYC and LA, who inspired us to the point we felt the need to write a book about it. Without you guys, none of this would have been possible.

Wink and Shotgun, Deb and Liv.

Flirtexting Notes